TRANSACTIONS

of the

American Philosophical Society

Held at Philadelphia for Promoting Useful Knowledge

VOLUME 77, Part 1, 1987

A Late Triassic
Footprint Fauna
From the Culpeper Basin
Northern Virginia (U.S.A.)

ROBERT E. WEEMS
U.S. Geological Survey

THE AMERICAN PHILOSOPHICAL SOCIETY

Independence Square, Philadelphia

1987

Library of Congress Catalog
Card Number 84-45904
International Standard Book Number 0-87169-771-8
US ISSN 0065-9746

CONTENTS

ILLUSTRATIONS

ABSTRACT

Thousands of footprint impressions, probably of Norian age, have been discovered on a single bedding plane in a quarry in the Culpeper Basin of northern Virginia. About 830 tracks on this bedding surface, contained in 32 recognizable trackways, were studied in detail. The other tracks were too obscure for meaningful analysis. These tracks are referred to six archosaurs, all probably dinosaurs: *Agrestipus hottoni* n. sp., *Anchisauripus parallelus*, *Apatichnus minor*, *Eubrontes* sp., *Gregaripus bairdi* n. sp., and *Grallator*(?) sp. These species are considered to represent four carnivores (*Anchisauripus, Apatichnus, Eubrontes, Grallator*[?]) and two herbivores (*Agrestipus, Gregaripus*) based on their foot shape, claw development, and the osteologically determined dietary preferences of animals in the group to which each trackmaker is assigned.

Herbivores greatly outnumber carnivores, and small herbivores are more abundant than large ones. The order of appearance of these trackmakers suggests that smaller and less agile species preferred soft ground, whereas larger carnivorous forms preferred a firmer substrate. From the measured print sizes, stride lengths, and pace angles, it was possible to estimate the hip height, body length, Froude numbers, and speed of each trackmaker. These results suggest that, in the environment in which these tracks were preserved, *Agrestipus* was a slow walker, *Anchisauripus, Apatichnus,* and *Eubrontes* were walkers to trotters, and *Gregaripus* and *Grallator*(?) were rapid runners.

Grallator(?), probably a coelurosaur, achieved high speeds and made sudden turns without skidding on a very soft substrate. These characteristics suggest that this animal was not entirely dependent on the substrate for support and stability; therefore it may have been feathered. Feathers would have provided enough lift and maneuverability, through use of feathered forelegs for banking against the air at turns, to account for the observed agility of this form on very soft ground. *Gregaripus* was probably an ornithischian and probably gregarious. *Agrestipus* was probably a sauropod. *Anchisauripus, Apatichnus,* and the species of *Eubrontes* reported here were probably carnosaurs.

Comparisons of relative stride length, striding rates, speeds, and pace angles of each species show that each had its own distinctive mode of locomotion. The consistently high activity levels of *Gregaripus* and *Grallator*(?) suggest that at least these forms were functional endotherms. The directions and relative order of passage of these species show a systematic shift from south-southeasterly (early arrivals) to easterly (late arrivals). This change of direction is interpreted as reflecting the existence at that time of an evaporating body of water north of this locality. As the edge of the water receded northward, the animals shifted their direction of travel to remain near it.

INTRODUCTION

The search for fossil bones of amphibians, reptiles, and mammals in sediments of the Upper Triassic/Lower Jurassic Newark Supergroup in eastern North America has continued sporadically for almost 200 years. Despite diligent efforts by many workers, the total accumulated osteological fauna is still sparse compared with that from many beds of comparable age elsewhere in the world. Thick sections of sediments may yield little or nothing diagnostic, but one or a few horizons locally may produce abundantly (e.g., Emmons, 1857; Lull, 1953; Colbert, 1965; Olsen and others, 1978; Weems, 1980).

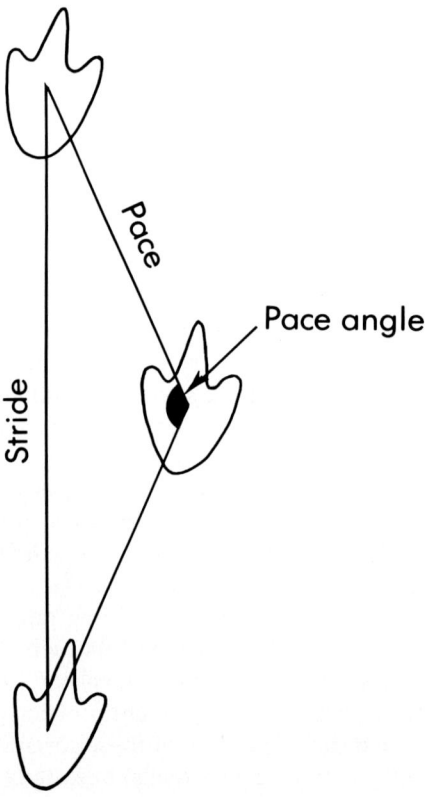

FIG. 1. Diagram showing the method used in this study to measure stride, pace, and pace angle (diagram modified from Sargent, 1975).

In contrast to this sparsity of bony remains, abundant footprints have been found in many of the Newark basins since Pliny Moody first mentioned one from Massachusetts in 1802 (cited in Lull, 1953). The large collections of footprints made from the Connecticut Valley area were first fully summarized by Hitchcock (1858) and were revised by Lull (1953). Footprints from the Newark basin have been described by Bock (1952) and Baird (1954, 1957). Photographs of footprints from high in the Jurassic column of the Culpeper basin of Northern Virginia were illustrated in Roberts (1928). These have been re-evaluated recently by Pannell (1985). A few footprints were figured by Shaler and Woodworth (1899) from the Triassic Richmond basin of Virginia, and footprints have been described from the Triassic Danville/Dan River basin by Olsen and others (1978).

Despite their relatively greater abundance, however, footprint faunas offer several difficulties for the taxonomist. In some living groups, especially birds, the feet of quite dissimilar animals may look very much alike and may make very similar trackways (Sargent, 1975). Conversely, sexual dimorphism, age variations, behavioral variations (for example, walking vs. running) or substrate variation (firm vs. soft) all may make the footprints of animals belonging to the same species look very different. Therefore, taxonomic conclusions must be drawn cautiously and preferably only when a large sample of tracks is available. Because trackway "species" probably reflect osteological genera (Baird, 1957), the application of the same generic and even species names to similar tracks occurring in both Late Triassic and Early Jurassic beds is not considered here to reflect inordinate time ranges for these taxa. Some of the names used in this report have had a varied history of higher taxonomic placement. The volume by Haubold (1971), which constitutes the latest overview of paleoichnology, summarizes much of the history of these controversial taxa. The pace, stride, and pace-angle terminology of Baird as adopted by Sargent (1975) is used here (fig. 1).

GEOLOGIC SETTING

The stratigraphic horizon of the footprints in this study is low in the rock column of the Culpeper basin. This horizon, and the horizon near Dulles Airport which yielded bones of *Rutiodon* (Weems, 1979), appear to be in the same lithostratigraphic unit. This has been called the Bull Run Shale by Roberts (1928), the Balls Bluff Siltstone by Lee (1977), and the Bull Run Formation by Lindholm (1979). Cornet (1977) and Cornet and others (1973) determined that all sediments above the lowest lava flow in the Culpeper basin are Jurassic in age, whereas most sediments below the flows are Norian in age. Cornet (oral communication, 1980) has processed shale samples collected at a horizon slightly above the Culpeper locality (fig. 2), provided by Albert J. Froelich of the U.S. Geological Survey. These samples contained middle Norian palynomorphs, which implies that the Culpeper track locality is no younger than middle Norian in age. No Carnian palynofloras have been documented from the Culpeper basin, but the undated basal exposed beds of the basin (Manassas Sandstone) may be as old as the upper part of that stage. Those basal beds appear to lie below the footprint horizon.

The footprints were found in the Culpeper Crushed Stone Quarry east of Culpeper, a locality briefly described in Kingston and others (1976). The footprints all came from a single bedding plane, now largely obliterated, within a cyclic sequence of thermally metamorphosed sandstones, siltstones, and shales (Plate 1A). Elsewhere in the Culpeper basin, many beds in similar sequences are red in color, so the pervasively gray color of all of these beds in the quarry partly may be secondary (metamorphic) in origin. However, because many of the footprints are very deep and indistinct and there are mudcracks (see Plate 2A), and because oolitic limestone and well-bedded siltstone or shale have been reported from the beds immediately overlying the footprints in the quarry (Young and Edmundson, 1954; Carozzi, 1964), at least some of the beds in the vicinity of the footprint horizon probably formed in standing water. Such beds may have been gray originally.

Fortunately, the print-bearing bedding plane separated readily in blasting and was a favored separation zone for quarrying. This resulted (fig. 3) in the clearing of an area several acres in extent along this plane. Literally thousands of prints were present, but about 830 were recognizable as belonging to discrete trackways. These prints were selectively studied. None

4

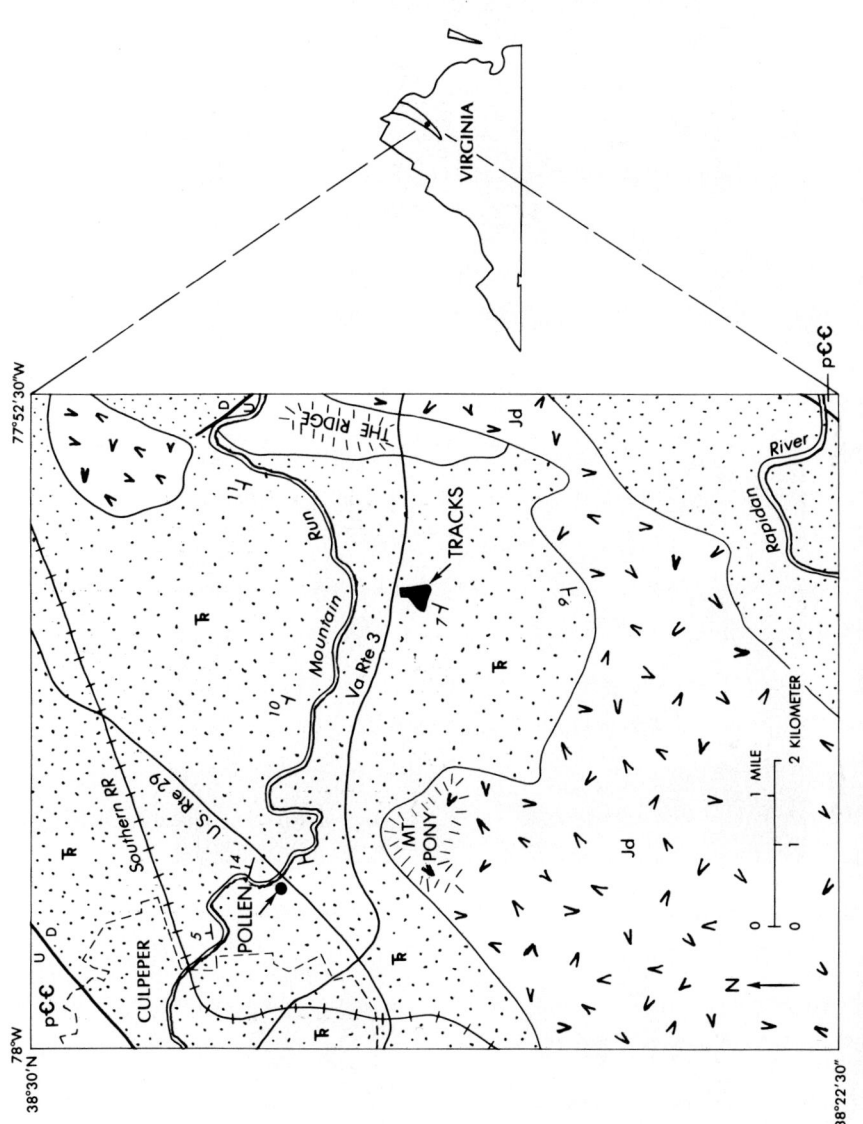

FIG. 2. Map of the area around the Culpeper footprint locality (labeled "tracks") and the nearest palynomorph-producing locality (labeled "pollen"). p€€, Precambrian to Cambrian rocks northwest and southeast of the Culpeper basin; Tr, Triassic-age sediments filling the Culpeper basin, Jd, Jurassic-age diabase injected into the Triassic sediment stack; U/D, fault with upthrown (U) and downthrown (D) sides indicated. Geology mapped by K. Y. Lee, U.S. Geological Survey.

A

B

C

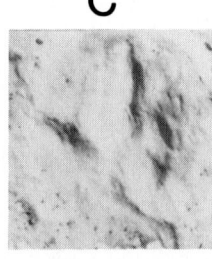

D

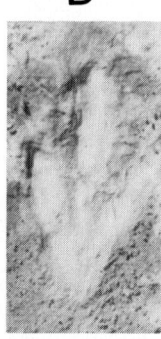

E
F

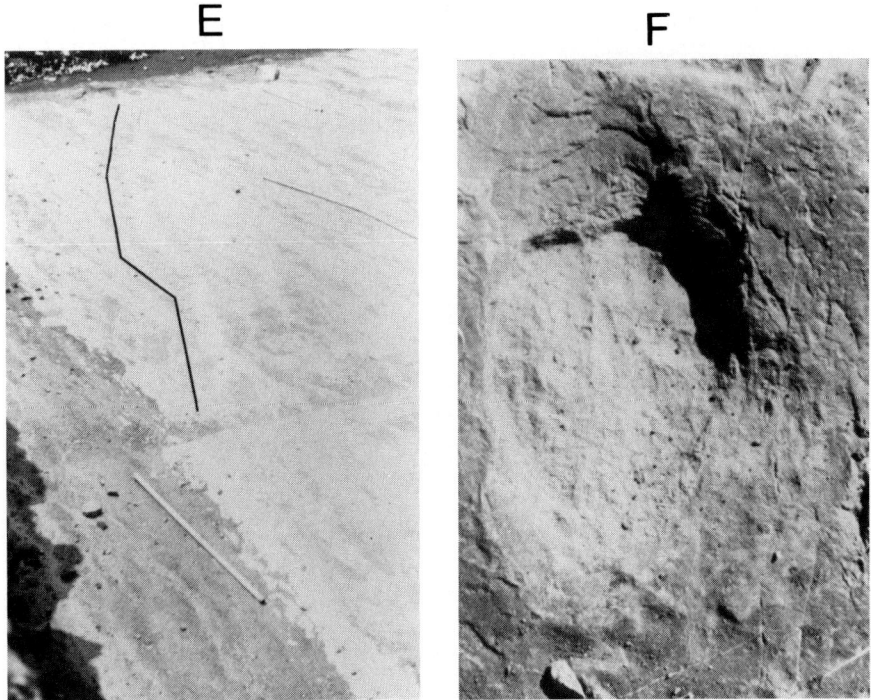

PLATE 1. A. View of quarry, showing the bedding surface on which footprints were observed (level above machinery). B. Two prints of *Eubrontes* sp., toes pointing right. C. Print of *Apatichnus minor*, toes pointing up and strongly divergent digit to left. D. Print of *Anchisauripus parallelus*, toes pointing up. E. Trackway of *Grallator*(?) sp. (GR3), viewed from above. The cleared trackway is light in appearance (to right of black line) and shows the zigzag pattern of travel. F. Footprint of *Agrestipus hottoni*, toes pointing up, from trackway AG7. Toe print to right is shadowed.

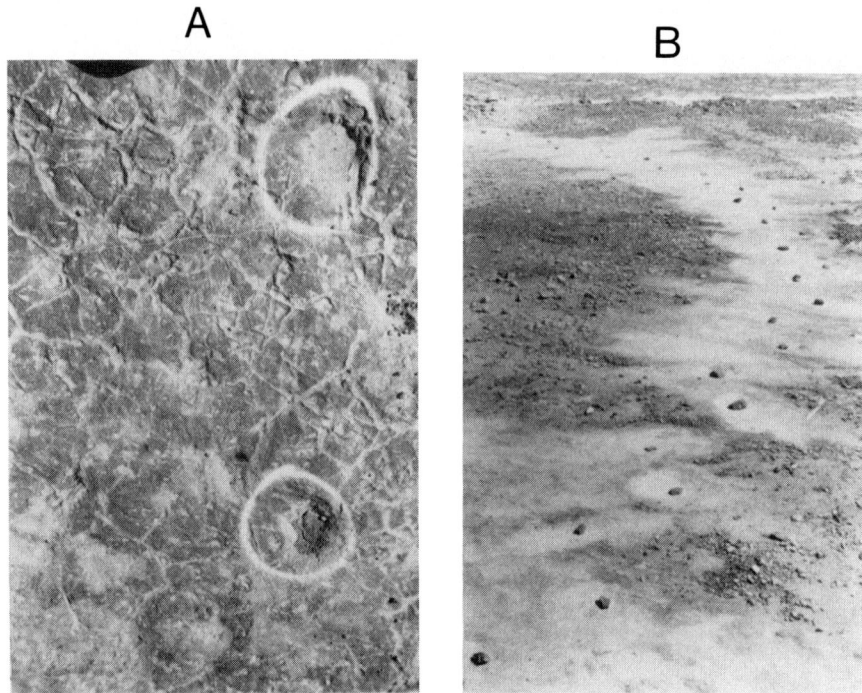

PLATE 2. A. Two footprints of *Gregaripus bairdi* pointing to the top of the page (circled), and mudcracks. B. Tracks of *Agrestipus hottoni* (AG8), marked by small stones. C. Print of *Agrestipus hottoni*, toes pointing up. D. Print of *Gregaripus bairdi*, toes pointing up. Quarter for scale. E. Tracks of *Gregaripus bairdi* (GR13), marked by small stones. F. Tracks of *Agrestipus hottoni* (AG7), marked by small stones. The white lines are cracks filled by calcite and other minerals.

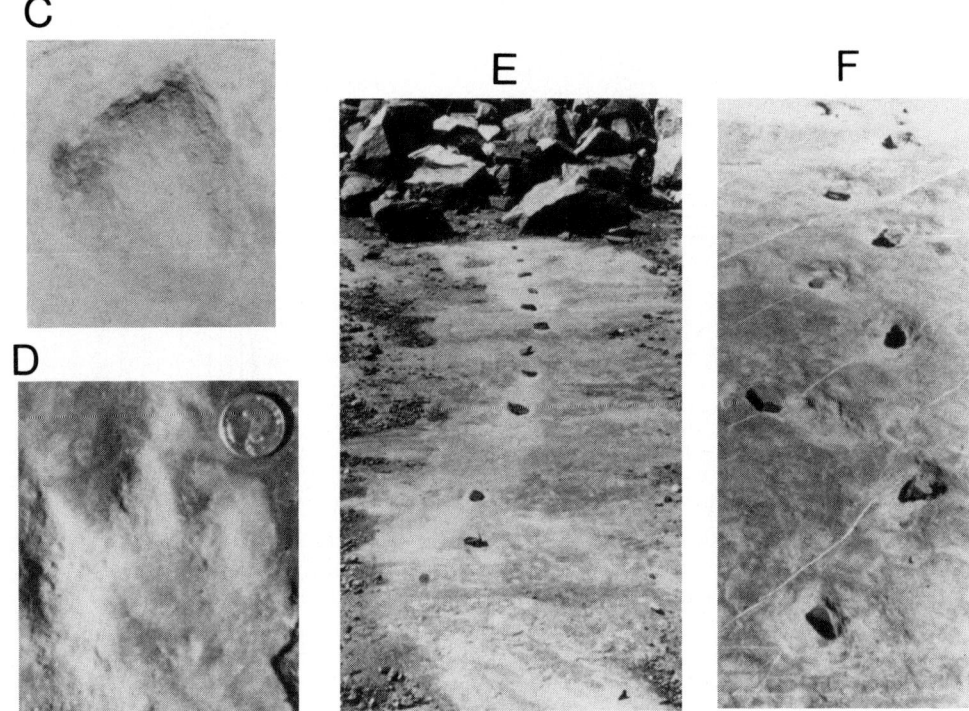

PLATE 2. (*Continued*)

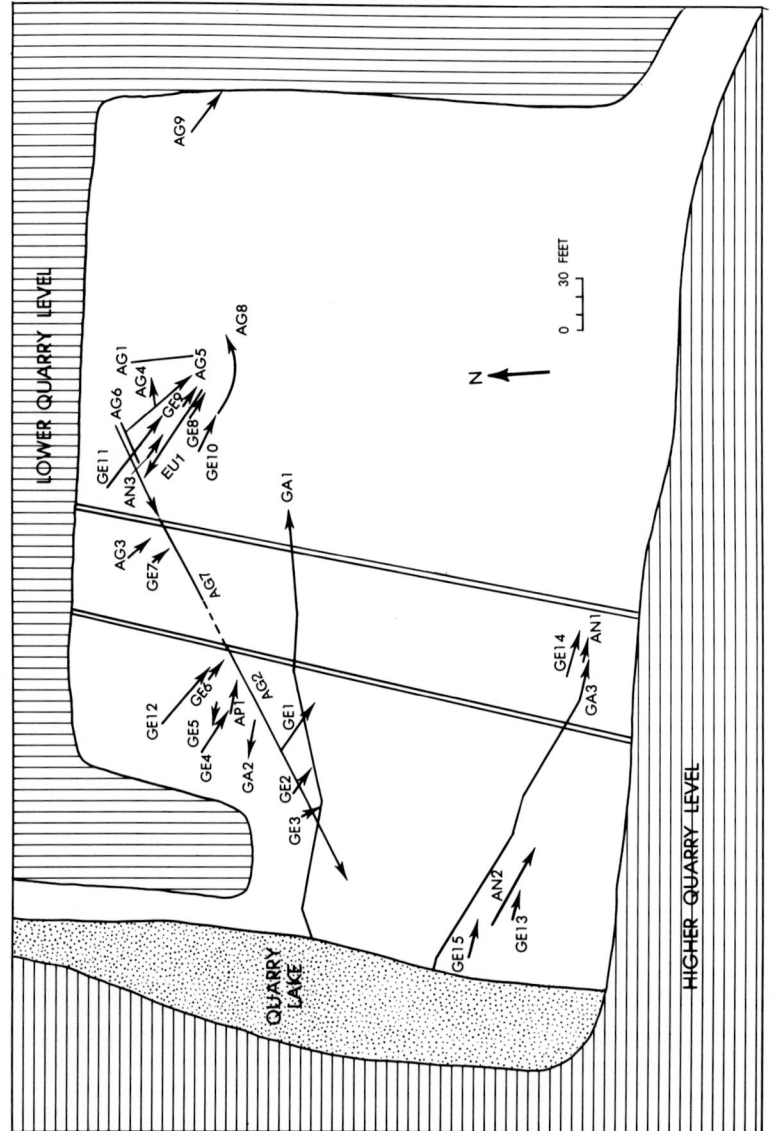

FIG. 3. Diagram of the footprint-bearing surface in the Culpeper Crushed Stone Quarry, showing the location, length, and orientation of the major trackways. The lake is modern. Patchy distribution of trackways has resulted from a combination of original patchiness, location of rubble in the quarry, plus selective bias due to areas deliberately cleared for this study. AG = *Agrestipus hottoni*, AN = *Anchisauripus parallelus*, AP = *Apatichnus minor*, EU = *Eubrontes* sp., GE = *Gregaripus bairdi*; GA = *Grallator(?)* sp. The two sets of parallel lines locate small faults offsetting the surface of the track-bearing bedding plane.

of them is of very high quality, either because the original surface was soupy, or because the exposed surface really consists of underprints beneath evaporites later ripped up by water currents to form the immediately over-lying oolites reported by Young and Edmundson (1954) and Carozzi (1964), or because of a combination of both factors.

TAXONOMIC ANALYSIS

Out of the thousands of more or less obscured footprints on the single track-bearing bedding plane in the Culpeper Crushed Stone Quarry, about 830 were identified as belonging to 32 separate trackways. These were made by at least six different kinds of animals (fig. 4), all probably dinosaurs. Five were recognized by the size and shape of the tracks, whereas one (*Grallator*(?) sp.) was recognizable mostly by a

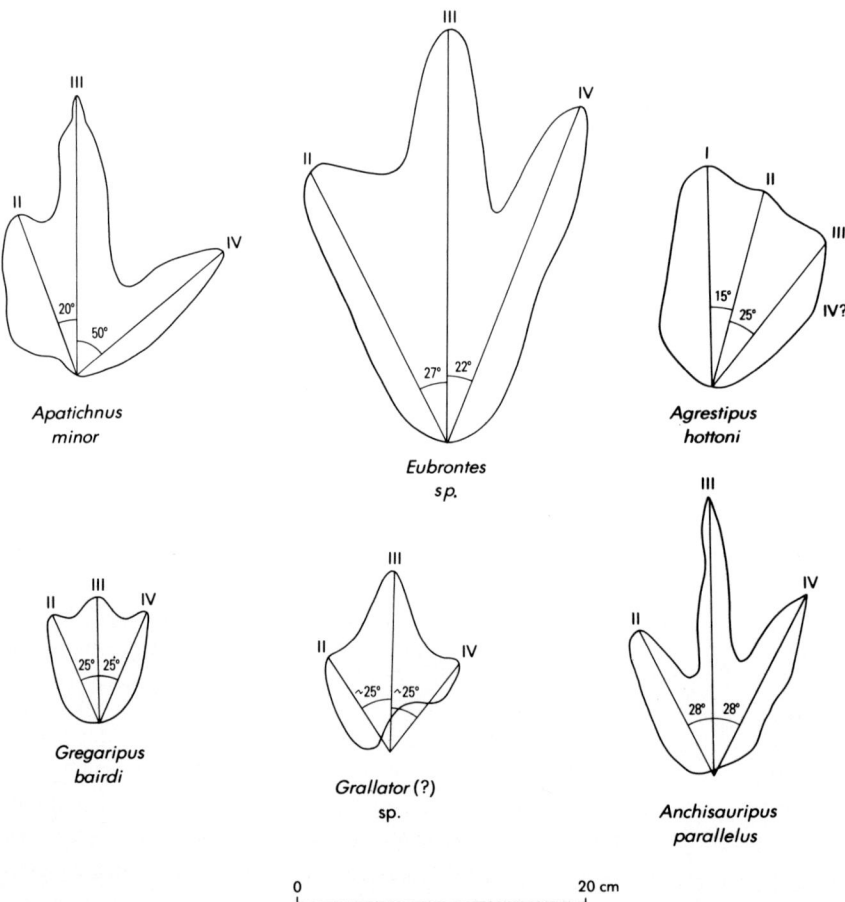

FIG. 4. Outline drawings of typical right footprints made by the six trackmakers at the Culpeper Crushed Stone Quarry.

distinctly long spacing between tracks and a zigzagging trackway pattern. These prints have been taxonomically assigned as follows:

Agrestipus hottoni gen. et sp. nov.

Holotype: United States National Museum (USNM) 358653, a set of four consecutive prints from trackway AG-2 (see Appendix 1) recorded as plaster molds.

Diagnosis of genus: A moderate-size pes impression (length about 15 cm, width about 11 cm), with three broad, blunt digits (I–III). Digit I is longest, the second and third successively shorter. Rarely, impressions of digit IV may be faintly present well behind digit III. The angle between I and II is about 15°, between II and III, about 25°. Print slightly longer than wide. Average pace known to range from 33 to 45 cm, average stride from 54 to 78 cm, and average pace angle from 92° to 139°. The trackway characteristics are distinctive (wide, short pace, low pace angle), and this trackway could be readily recognized even when individual prints were very obscure (fig. 5).

Referred specimens: USNM 358662 from trackway AG-4, USNM 358664 from trackway AG-7 (see Appendix 1).

Diagnosis of species: As the genus is currently monotypic, the diagnosis is as for the genus.

Derivation of name: Genus—for the Latin equivalent of the field name "clumsy," here "clumsy foot," in allusion to its short pace and low pace angle. Species—for Nicholas Hotton III of the U.S. National Museum of Natural History.

Discussion: The trackmaker of *Agrestipus hottoni* (fig. 4, Plates 1F, 2B, 2E) produced prints ranging from indistinct circular "puddles" marked by concentric ripplelike rings to rather ducklike prints having at least three blunt toes at the end of the foot and occasional hints of a fourth somewhat shorter toe on the external border of the foot. Although many of the prints are indistinct, the foot in general size and shape is similar to that of *Brachycheirotherium parvum* (Baird, 1957), and the shape is also reminiscent of *Deuterotetrapous* sp. from Africa (Haubold, 1971, p. 97) and *Navahopus* from Arizona (Baird, 1980). *Agrestipus hottoni* is similar to all of these forms in that it has a persistently low pace angle, but it differs from all of them in that digit I is the longest. In *Brachycheirotherium* and *Navahopus* digit III is longest and in *Deuterotetrapous* digit II is longest. Usually only pes digits I–III are firmly impressed, but, rarely, a hint of pes digit IV shows up about one third of the way back of digit III toward the rear of the foot. The emphasis on digit I suggests that this animal had sauropod affinities.

Charig and others (1965) have argued that African *Deuterotetrapous* sp. prints of Ladinian or possibly Carnian age may have represented a true sauropod because the manus prints are present, but whether *obligate* bi-

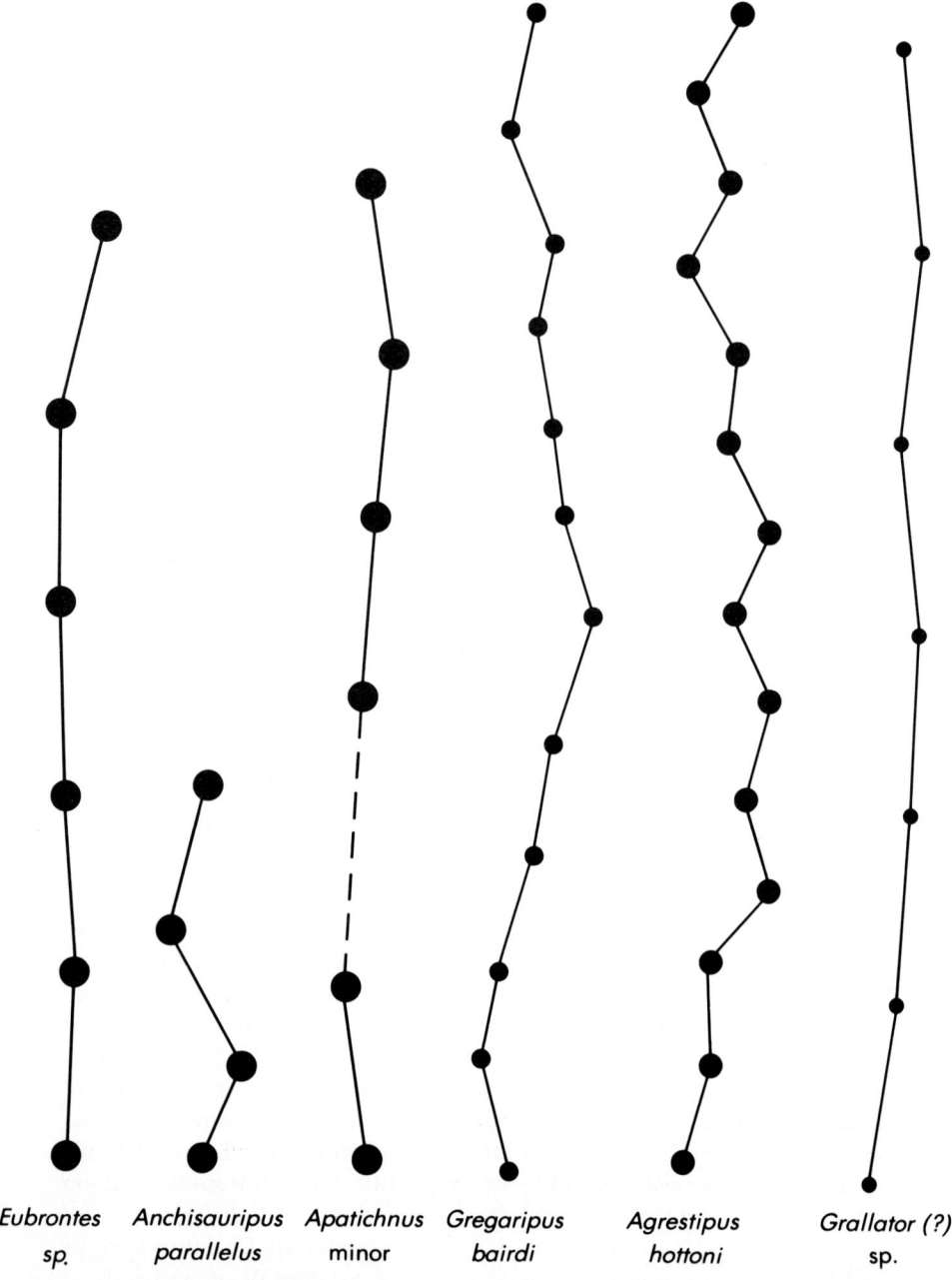

Eubrontes   Anchisauripus   Apatichnus   Gregaripus   Agrestipus   Grallator (?)
sp.   parallelus   minor   bairdi   hottoni   sp.

FIG. 5. Diagram showing typical trackway segments of the six types of tracks found at the Culpeper Crushed Stone Quarry. Circles reflect the relative sizes of the footprints except for those of *Grallator*(?), which mostly left only the impressions of the tips of its toes.

pedalism or quadrupedalism was present in either *Agrestipus* or *Deutero-tetrapous* is by no means clear. Moreover, at certain speeds quadrupedal animals step almost exactly on the prints made by their front feet and obliterate the manus prints, thereby creating a superficial but incorrect impression that the trackway was made by a bipedal animal. Thus it is not absolutely certain that *Agrestipus* was the bipedal animal that it appears to be.

For these reasons, there are no compelling reasons to bar *Agrestipus* from the sauropod lineage. On the basis of skeletal material, the sauropod lineage cannot be traced earlier than the Early Jurassic (Jain and others, 1977), but what is known of these early forms suggests that already in the Early Jurassic they were much like the later, well-known Late Jurassic and Cretaceous forms. Thus the relative proportions and emphasis of the toes, very much like those in later sauropods, offer positive reasons to include *Agrestipus* among this group rather than among the prosauropods, from which it was probably derived.

Anchisauripus parallelus (E. Hitchcock 1865)

Trackways of this species include some of the clearest prints found (fig. 4, Plate 1C). The rearwardly directed hallux print, a unique trait of *Anchisauripus* among Newark ichnogenera, is sometimes clearly discernible. The foot is relatively narrow and long, with the lateral toes very unequal in length. Only *A. exsertus* and *A. parallelus* show such a great difference in the length of digits II and IV. The toes are too narrow, and the hallux too rearwardly rotated, to be that of *A. exsertus;* therefore this track is assigned to *A. parallelus.* Tracks of this shape were made by either coelurosaurian or carnosaurian dinosaurs (Baird, 1980). Because the locomotor characteristics are most comparable to those of *Apatichnus* (discussed previously) and *Eubrontes* (discussed later and see figs. 10–15), *Anchisauripus* tentatively is considered to be a small carnosaur.

Trackway AN1, possibly an extension of AN2, has an unexpectedly short and irregular gait. These four tracks are on a well-preserved part of the bedding surface, yet there is no trace of tracks immediately ahead or behind these four. Perhaps this animal hit a soft spot on otherwise hard ground, causing it to stumble. The lack of other deeply impressed *Anchisauripus* trackways like this, and the lack of a longer trackway for this particular animal, both make a local soft spot the most likely explanation for its peculiarities.

Referred specimens: USNM 358665 from trackway AN-1, USNM 358661 from trackway AN-3 (see Appendix 1).

Apatichnus minor (E. Hitchcock 1858)

This trackway (fig. 4) includes very distinctive prints, on which digit IV is strongly rotated away from digit III at an angle of about 50°. The strong divergence between digits III and IV is quite distinctive of *Apatichnus*, and for this reason this trackway can be readily assigned to that genus even though it originally was described from the Jurassic part of the Newark column. Digit I was not discernible, but it is only faintly impressed in even the best of *Apatichnus* prints. In size, this trackway is very similar to that of *A. minor* and is much larger than *A. circumagens*. As is the case in *A. minor*, and unlike *A. circumagens*, there is no trace of the manus. This track-maker previously has been associated with ornithischian (Lull, 1953) and prosauropod (Haubold, 1971) dinosaurs, but its gait (discussed later) is most like that of *Anchisauripus*. Therefore, *Apatichnus* here is associated with the carnosaurs, though reference to the coelurosaurs also may be plausible.

Referred specimen: USNM 358665 from trackway AP-1 (see Appendix 1).

Eubrontes sp.

Tracks of one trackmaker (fig. 4, Plate 1B), the largest found in the Culpeper quarry, are very similar in size and shape to tracks assigned to this genus, though the detail is too poor to allow specific identification. Among Newark trackmakers, only *Apatichnus, Otozoum, Gigandipus, Eubrontes, Anchisauripus,* and possibly *Grallator* attain sizes comparable to this print. *Otozoum* has four instead of three large toes, with the third less elongate than in this Culpeper species, while *Grallator* and *Anchisauripus* both have a much more elongate digit III. Although one digit in this form (II) is somewhat divergent from the other two, it is not nearly so divergent as the fourth digit of *Apatichnus*. Therefore it cannot be assigned to that form, even if the prints are improperly sorted as to which are the left and right sets. *Gigandipus*, although somewhat similar in gross proportions, typically shows a very distinct hallux impression not evident on any of these prints. Only the present definition of *Eubrontes* fully encompasses all of the observable characteristics of this form. The tracks of this very large animal are faint, suggesting that it was the last animal to leave prints on this particular bedding plane. Its pace was about 85 cm, with a pace angle of about 174°.

It seems possible that the ichnogenus *Eubrontes* may encompass tracks made by two very different types of trackmakers which are only distantly related. Although large-headed carnivores such as *Postosuchus* have a foot which corresponds closely in proportions to many of the tracks called *Eubrontes* (Chatterjee, 1985), structural rear foot and limb differences between Triassic carnivorous theropod dinosaurs and some herbivorous prosauropods may not be very great (Charig and others, 1965). Therefore a general similarity in shape between *Eubrontes* prints and those of carnosaurs by itself does not constitute a compelling argument for assuming that all of these forms represent carnosaurs. Although many prosauropods (such as the skeletal genera *Lufengosaurus* [Young, 1941] and *Ammosaurus* [Baird, 1980] and the trackway *Navahopus* [Baird, 1980]) still retain a well-developed digit I and represent a more primitive, functionally four-toed grade of foot development than is seen here, *Plateosaurus* (a small-headed herbivore) seems to have advanced beyond this level and become functionally tridactyl (Huene, 1908, 1926).

It seems quite likely that *Eubrontes giganteus,* the genotype of *Eubrontes,* could have been made by a descendant of *Plateosaurus.* While *E. giganteus* often has been considered to be a carnivore (Ostrom, 1971, among others), its regional abundance and relatively very large size in conjuction with its apparently gregarious nature suggest otherwise. It is true that modern medium-sized carnivores such as lions and hyenas hunt in packs in order to collectively kill larger prey and then take shares. But when the largest animal in a community (such as *E. giganteus*) is a carnivore, it could not be expected to gain individual advantage by pooling efforts to hunt smaller forms because the amount of meat available to each pack member from a kill would be less than to each one individually. Lull (1953, p. 178) long ago noted that *Eubrontes giganteus* lacks the trenchant, raptatorial claws typical of carnivores, and for this reason doubted that it could have been wholly carnivorous. Additionally, it should be noted that *E. giganteus* of the Early Jurassic has a pace angle of only about 160°. This is higher (more advanced) than that of the most efficiently walking Late Triassic *Agrestipus* (probably a sauropod), but lower (less advanced) than that in Late Triassic theropods such as *Anchisauripus* and *Grallator.* For all of these reasons, the trackmaker which produced *Eubrontes giganteus* is suspected to be a progressive tridactyl, more-or-less herbivorous prosauropod rather than a tridactyl carnivore.

The trackmaker which made the tracks here called *Eubrontes* sp. had a pace angle of about 174°, which is quite comparable to that of Late Triassic carnivorous theropods like *Anchisauripus* and *Grallator* but much higher than that of the Early Jurassic, possibly prosauropodian, *Eubrontes giganteus.* For this reason, *Eubrontes* sp. is considered to have carnosaur rather than prosauropod affinities. Because the size, relative foot proportions, and short digit III on the tracks of this animal are closely comparable to the pes of the recently described Late Triassic skeletal genus *Postosuchus* (family Poposauridae), and because *Postosuchus* was almost the same length (4 m) as the estimated length of the animal that made these prints (3.9 m), it is tentatively associated with that genus. *Postosuchus* shows obvious carnosaurian (probably tyrannosaurid) tendencies (Chatterjee, 1985).

Referred specimen: USNM 358656 from trackway EU-1.

Grallator(?) sp.

This trackmaker (fig. 4, Plate 1D) left no distinct individual prints, but nevertheless it can be recognized readily by its long pace (87 cm) relative to its small apparent print size (12 cm). The overall shape of the foot is most suggestive of three toes, the middle one by far being the longest. The best preserved print is about 12 cm long, 9 cm wide, and has about a 25°

divergence between II and III and a 25° divergence between III and IV. Informally called "Speedy" during fieldwork, this animal is assumed to represent a small coelurosaur, and for that reason is placed tentatively in the genus *Grallator*. Its proportions are comparable to those of *G. cuneatus* Hitchcock. Placement in *Anchisauripus* would also be plausible, though all well-known species in that genus seem to be two or more times larger than the size indicated by these prints.

Referred specimens: USNM 358654 from trackway GA-1, USNM 358657 from trackway GA-2 (see Appendix 1).

Gregaripus bairdi gen. et sp. nov.

Holotype: USNM 358651, an isolated print, the only original recovered from the pit. Sets of four consecutive prints from trackway GE-1 preserved as plaster molds (USNM 358652) are designated a paratype.

Referred specimens: USNM 358659 from trackway GE-11, USNM 358660 from trackway GE-2 (see Appendix 1).

Diagnosis of genus: A consistently small footprint with three broad, blunt digits, bearing only blunt claws, foot slightly longer than wide. Digits II and IV arranged symmetrically on either side of digit III. Track length 8 to 10 cm. Average pace known to range from 44 to 58 cm; average stride, from 86 to 113 cm; and average pace angle, from 131° to 176°.

Diagnosis of species: As the genus is currently monotypic, the diagnosis is as for the genus.

Derivation of name: Genus—Latin for "sociable foot," from its tendency to occur in flocks. Species—for Donald Baird of Princeton University.

Discussion: This trackmaker (fig. 4, Plates 2A, 2C, 2D) had three nearly equally long toes. Though *Anomoepus isodactylus,* as described by Lull (1953), had a small fourth toe, it has an otherwise similarly shaped print. However, the print of *A. isodactylus* is about twice the size of the largest of the Culpeper prints, and this Culpeper track type, more than any other, is clearly impressed on the bedding surface. Therefore, the difference in size and the lack of any evidence of a fourth toe among hundreds of prints is considered significant. This print is probably that of the same animal that Baird (1957) described as *"Genus incertum"* from New Jersey and that was illustrated from Pennsylvania without a name in Wanner (1926). Because of its blunt toes, abundance, and its probably gregarious nature (discussed later), this animal was probably herbivorous. The size and shape of the foot are similar to those of the foot of *Heterodontosaurus tucki* (Luca and others, 1976). Therefore, this trackmaker is considered to have been a small ornithischian dinosaur.

DISCUSSION

The trackmakers at this locality were all contemporaries. The bedding plane that represents the sediment surface across which they wandered was under water, and thus inaccessible, immediately prior to the time that the trackmakers wandered across it. Probably it was soft enough to accumulate the observed prints for only a few days before it dried out too much to record the passage of any more animals over it. Thus, we see at this locality traces of part of a true biological community. The generally poor preservation on this surface precludes recognition of any small trackmakers, so the record here is biased toward larger animals. Yet, with that understanding, several conclusions can be made concerning animals probably ranging in length from 1 to 4 m. These conclusions are summarized as follows:

Relative Numbers of Carnivores and Herbivores

Figure 6 shows the relative numbers of each type of trackway. *Agrestipus* and *Gregaripus* are here presumed to be herbivores, and *Grallator*(?), *Anchisauripus*, *Apatichnus*, and *Eubrontes* to be carnivores. The presumed roles of *Grallator*(?) and *Anchisauripus* as carnivores and *Agrestipus* and *Gregaripus* as herbivores would not seem to be controversial. Although Lull (1953) considered *Apatichnus* to be an herbivore, its trackway characteristics (discussed later) are most similar to *Anchisauripus*, long accepted as a carnivore. Therefore, *Apatichnus* here is considered to be a carnivore. The reasons for calling *Eubrontes* sp. a carnivore were discussed previously.

If these various ecological roles are accepted, small herbivores are most abundant, large herbivores are second most abundant, and carnivores are least abundant. Among carnivores, the small *Grallator*(?) outnumber each type of large carnivore, but the large carnivores collectively outnumber *Grallator*(?). This violates the food chain concept that large carnivores are among the rarest forms in any community. The apparent overabundance of large carnivores may have resulted from collecting bias (because the carnivore tracks are relatively large and are diverse taxonomically) or because large carnivores were lurking in exceptional numbers in this vicinity in expectation of the imminent arrival of the herbivores. Except for the overabundance of large carnivores, however, the relative abundances of each track type support the ecological roles presumed for each type of trackmaker.

Although the smallest carnivore *Grallator*(?) probably was too small to attack any of the observed herbivores, the ratio of predators to prey still

19

is only about 1:6. Relative size considerations suggest that *Anchisauripus* and *Apatichnus* may have been capable of killing *Gregaripus,* while *Eubrontes* sp. might have bested either *Gregaripus* or *Agrestipus* on firm ground.

Substrate Preferences and Order of Appearance

The degree to which the trackways were impressed into the drying mud (fig. 7) should vary according to the relative weights and speeds of the animals. But all the lightly impressed tracks are of the largest animals (*Anchisauripus, Apatichnus,* and *Eubrontes*). Because of the size of these animals, they should have produced deeper tracks than the small animals if they had passed through at the same time or earlier. Thus, they must have been the last animals to pass across this surface. Deeper tracks are separable into those whose form is distinct (*Gregaripus*) and into those whose form is obscure (*Agrestipus, Grallator*[?]). *Gregaripus* tracks are moderately to shallowly impressed, but their outlines are well formed. Some *Agrestipus* tracks approach this quality, though none are quite so clear. A few *Agrestipus* tracks show evidence of toe shape, so they are partly included in the moderately impressed range, because an animal of this size would not leave tracks quite as distinct as the lighter *Gregaripus* in the same consistency of mud. However, most *Agrestipus* trackways show almost no details, though their distinctive print size, pace length, and pace angle still characterize them. Although *Grallator*(?) left deeper tracks than any of the other trackmakers, even though it was a much smaller animal than *Agrestipus,* its pace length relative to its size indicates that it was moving at high speed

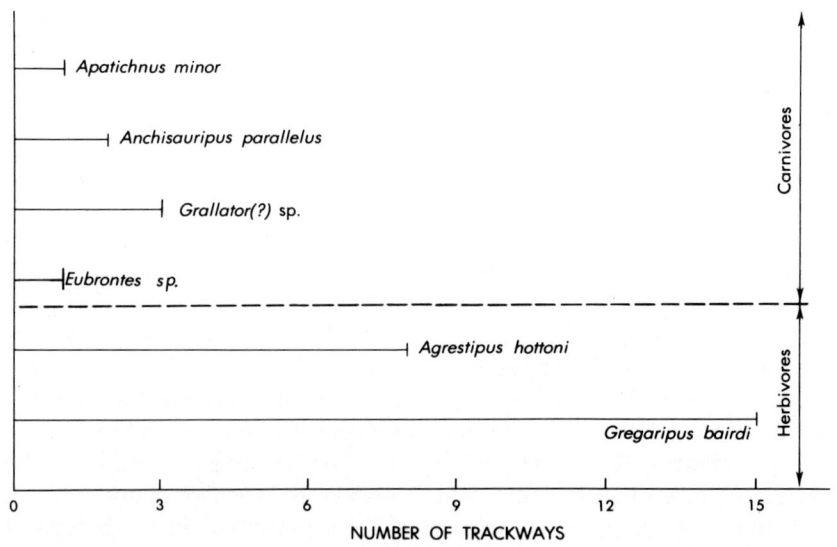

FIG. 6. Bar graph showing relative abundances of each kind of trackmaker at the Culpeper Crushed Stone Quarry locality. Note the preponderance of presumed herbivores in this sample.

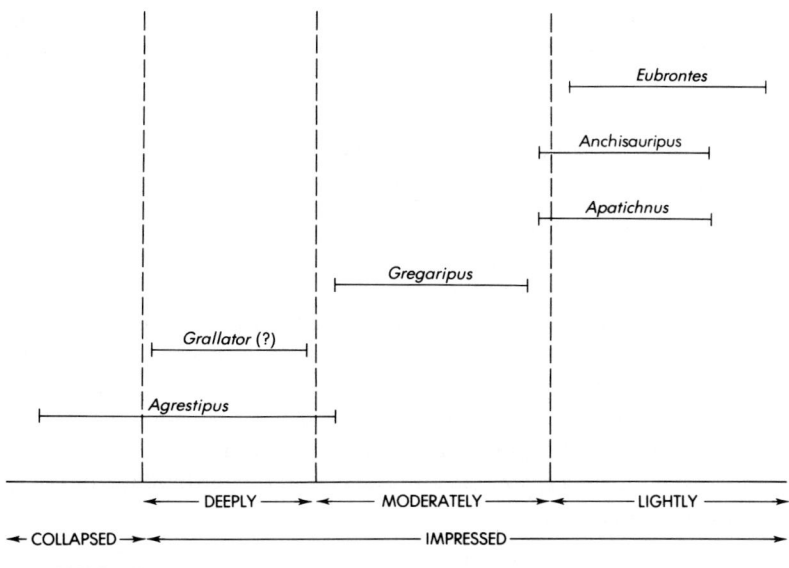

FIG. 7. Diagram showing the relative degree to which different trails are impressed at the Culpeper Crushed Stone Quarry. Lightly impressed trails have relief of about 0.5 cm, moderately impressed trails have relief of about 0.5 to 2 cm, and deep trails have relief greater than about 2 cm. Collapsed prints have been refilled by mud and are represented only by concentric ringlike patterns having less than 0.5 cm of relief. The Y axis of this graph also roughly corresponds to time elapsed during drying, so that the relative order of appearance and disappearance of each genus can be read from left to right.

(fig. 5). It is likely that its resultant high momentum, and not its earlier passage, caused it to leave deeper tracks than *Agrestipus*. Some *Agrestipus* trackways show the prints as only concentric rings, indicating that the mud was so soft that it collapsed back into the print. In contrast, the *Grallator*(?) prints, though deep, show no sign of the mud collapsing back into them. This strongly indicates that *Agrestipus* preceded *Grallator*(?) across this bedding plane.

Therefore, as the mud became progressively firmer through water loss, the order of passage was probably *Agrestipus, Grallator*(?), *Gregaripus, Anchisauripus* with *Apatichnus,* and finally *Eubrontes* (fig. 7). The presence of *Grallator*(?) so early in this sequence is puzzling for reasons which will be discussed later. Even so, all of the animals came through in a sequence which suggests that the smaller herbivores and carnivores preferred softer ground than the large carnivores. Possibly on soft ground it was easier for smaller animals to avoid predation by the large carnivores, which would have needed sure footing successfully to overtake their prey.

The low Froude number and the estimated normal walking speed of *Agrestipus* (see Table 1), suggest that this animal was walking exceptionally slowly in this environment. Perhaps *Agrestipus* was foraging for food in shallow water, in addition to avoiding predation by the large and formidable carnivore, *Eubrontes* sp., which came by only when the ground was nearly firm.

Locomotor Rates and Patterns

When the pace angles of successive tracks are compared (fig. 8), the contrast in locomotor pattern is striking. Although *Eubrontes* sp., *Apatichnus minor*, and *Anchisauripus parallelus* are known only from short trail segments, the patterns agree quite closely with that of *Grallator*(?) sp. All have high pace angles (170°–180°) that are consistently maintained. The very long *Grallator*(?) trackway does show a slightly sinusoidal pattern, suggesting a tendency to fluctuate speed in a rhythmic fashion. Overall, however, all these animals appear to have been very efficient walkers and runners that held their legs directly under their bodies.

By using the formula developed by Alexander (1976) to relate hip height and stride length to speed:

$$v \simeq 0.25\, g^{0.5} \lambda^{1.67} h^{-1.17}$$

where v = velocity, g = acceleration due to gravity, λ = stride length, and h = hip height $\simeq$ 4 times the foot length, it can be calculated that *Eubrontes* stood about 1.1 m (4 ft) at the hips and was moving about 1.7 m/sec (3.7

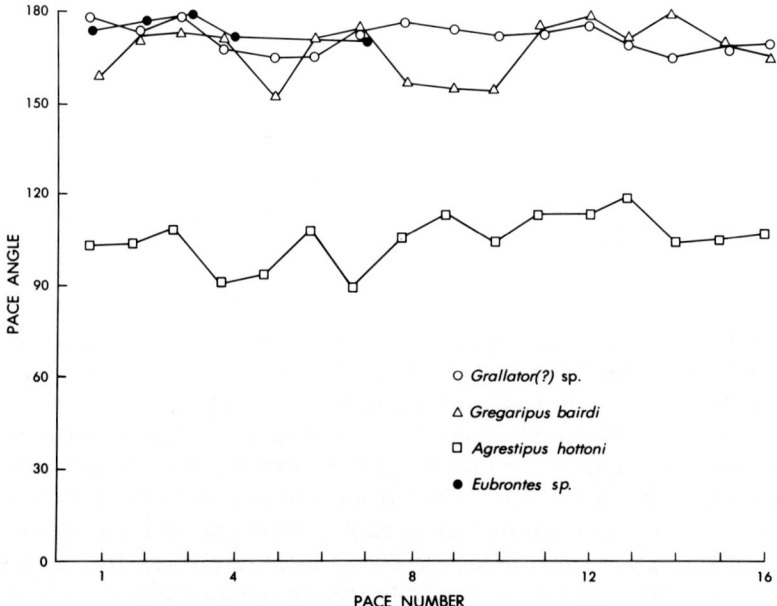

FIG. 8. Diagram showing successive pace angles in four continuous trackways. Black circles, *Eubrontes* sp.; open circles, *Grallator*(?) sp.; triangles, *Gregaripus bairdi*; squares, *Agrestipus hottoni*.

mph). Its relative stride length (see Table 1 and fig. 9 for terms) was 1.5. Its Froude number, which Alexander (1976) has suggested applies to the interaction of inertia and gravity in land vertebrate locomotion, can be determined by using the formula:

$$F = v^2/gh$$

where F is the Froude number, v is the estimated velocity, g is acceleration due to gravity, and h is the estimated hip height ($\simeq 4$ times the footprint length). For *Eubrontes* this value is 0.3, which suggests that it was walking and not running. Similarly, *Apatichnus* (estimated hip height 0.75 m) would have been moving about 2.3 m/sec, and *Anchisauripus* (estimated hip height 0.75 m) would have been moving about 1.4 to 2.8 m/sec. On the basis of their respective relative stride lengths (2.0 and 1.6–2.3) and derived Froude numbers (0.7 and 0.3–1.2), *Apatichnus* probably was moving at a fast walk or trot, whereas one *Anchisauripus* was trotting and the other (or the same?) was walking or stumbling (because it hit a patch of exceptionally soft substrate which recorded exceptionally deep footprints). Assuming a hip height to body length ratio of 1:3.5, *Eubrontes* was about 3.9 m (13 ft) long, and *Apatichnus* and *Anchisauripus* were about 2.6 m (8.5 ft) long. The similar size and comparable locomotor characteristics of *Anchisauripus* and *Apatichnus* suggest that they were closely related (figs. 10–15). Although *Eubrontes* has locomotor characteristics more similar to *Anchisauripus* and *Apatichnus* than to any of the other trackmakers (figs. 10–15), it does differ from them in detail (figs. 11–15) and thus represents a recognizably different kind of walker.

The print size of *Grallator*(?), indicates that it was an animal about 0.5 m (1.5 ft) high at the hips and 1.8 m (6 ft) long, about the dimensions of the coelurosaur *Coelophysis*. Considering that the stride of *Grallator*(?) was as long as 1.7 m, its speed is estimated at 4.6 m/sec (9.9 mph). Moreover, its relative stride length reached a maximum of 3.5, and its Froude number reached a maximum of 3.9, much higher than in any of the dinosaurs studied by Alexander. This means that this animal had a very long stride, compared with its leg length and proportionately related foot size, and it was not walking or trotting, as were some of the other larger track makers. The conclusion that this animal was running rapidly helps to explain the relatively great depth and poor clarity of these prints. However, this reasoning raises the question of how a small animal could move so rapidly and so surely across very soft and slippery mud. The size and shape of the foot would not seem to allow the animal to skip across mud, and the great depth of the tracks shows that the feet sank into the mud readily. Another puzzling feature of these trails is the way in which they zigzag (fig. 4, fig. 19, Plate 1E). The turns are abrupt, taking place at single tracks, and the individual tracks on the turns show no evidence of skidding or sliding. These observations seem self-contradictory for an animal that is entirely dependent on the ground for traction and support. If the animal bunched the toes on each foot as it lifted them from the ground, that would help

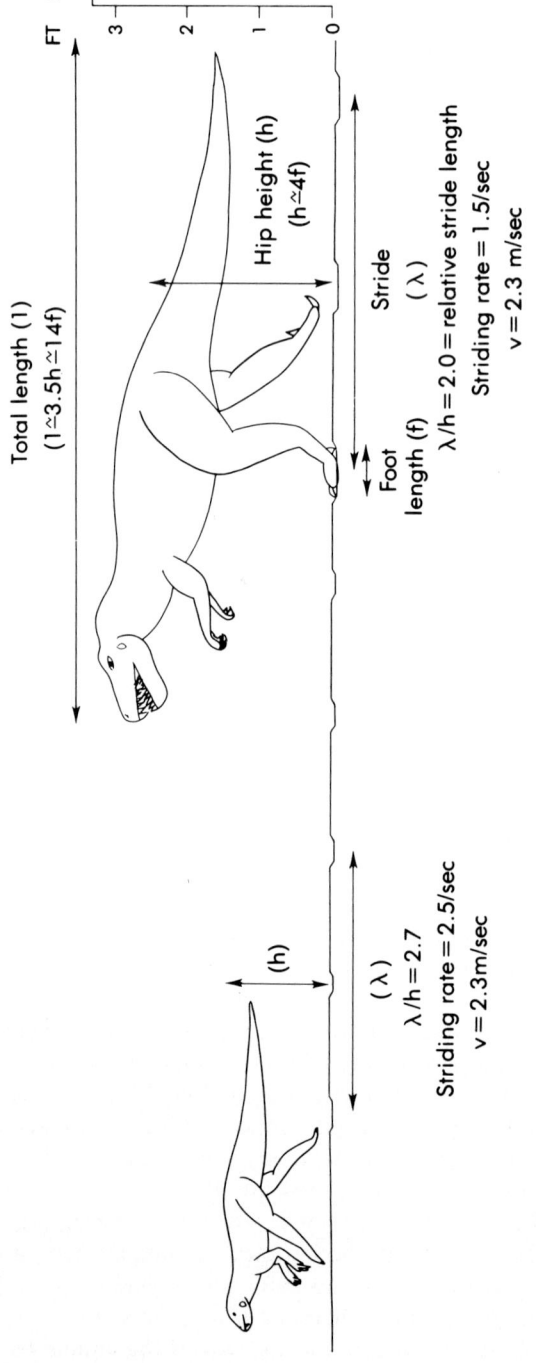

FIG. 9. Pictorial representation of possible appearance of *Anchisauripus* and *Gregaripus*, showing the relationships of relative stride length, stride, and hip height. Because *Gregaripus* was much smaller than *Anchisauripus*, it had to attain a much higher relative stride length, as well as a higher striding rate, in order to travel at the same velocity. As depicted here, a *Gregaripus* (modeled after *Heterodontosaurus*) is running at a velocity of 2.3 m/sec., while an *Anchisauripus* (portrayed as a generalized carnosaur) pursues at a fast walk that matches the velocity of the smaller animal.

to free each foot from the mud if it was sticky. That would have improved the animal's speed, but it would not have helped to stabilize the path of the animal as it turned.

One way to resolve this apparent paradox would be to suggest that this trackmaker was feathered. If this were so, then even though the animal probably could not fly, the lift from its feathers would make it light enough to keep from bogging down and would allow it to make sharp banks against the air and not the ground by maneuvering its feathered forelegs. Hollow bones, to which the name coelurosaur alludes, must have improved even further the lift that a body coat of feathers would have provided. Such an animal could be at home on marshy ground, and in such an environment it could easily escape from any half-mired large predator. The consistently high running speeds suggest that this animal had a high metabolic rate, and, if so, plumage would have helped to conserve body heat at night. For these reasons, even without flight, two major selective advantages would result from having plumage. Thus, the presence of feathers on some, or perhaps all, coelurosaurs by Norian time is entirely possible, and these animals may well have been fully endothermic, as has been suggested elsewhere (for example, Bakker, 1975).

It is also possible that only one lineage developed feathers, that which gave rise to birds. If so, then this *Grallator*(?) would be the earliest known protoavian and would push the avian lineage back to as early as the Norian, independent of the coelurosaurs. In either event, at least some coelurosaur-like creatures may well have been feathered and highly active at this early stage of dinosaur evolution.

The *Grallator*(?) trackways show a linear relationship between pace length and pace angle (fig. 10). This relationship reflects a tendency for the pace to lengthen as the animal went faster, permitting it to cover more ground. The larger carnivores collectively plot out along a paralleling trend, indicating they possessed a similar but slightly less efficient locomotor system than *Grallator*(?).

Gregaripus bairdi represents quite a different style of locomotion. Its pace angle is nearly as large as that in the previously discussed groups (fig. 10), but it varies over a much higher amplitude and has a much more irregular cycle than do the pace angles of *Grallator*(?) and *Eubrontes*. This animal, whose whole foot is clearly impressed, probably stood about 0.35 m (1.1 ft) at the hip and was about 1.2 m (3.7 ft) long. The size of *Gregaripus bairdi* matches almost perfectly the proportions of *Heterodontosaurus tucki*, as restored by Luca and others (1976), which stands 0.3 m (1 ft) at the hips and is about 1.0 m (3.3 ft) long. Use of Alexander's formula suggests that this animal was moving at speeds of 2.1 to 3.3 m/sec (4.7 to 7.4 mph), had a relative stride length of 2.5 to 3.3, and a Froude number of 1.3 to 3.1. This indicates that these animals also were active runners. The conclusion that all were running, and not all in the same direction, suggests that running may have been fairly normal locomotor behavior for these animals and not the result of a single mass stampede, as has been suggested

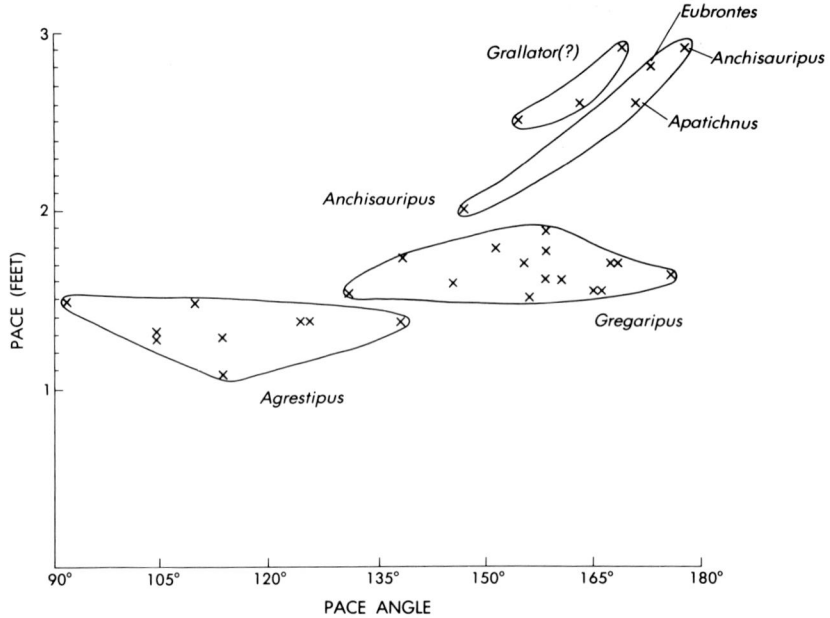

FIG. 10. Graph showing average pace of each trackway relative to its average pace angle. Fields containing all tracks of a particular species are outlined.

for trackways described from Australia (Thulborn and Wade, 1979). The fact that no carnosaur tracks were observed that are pressed more deeply than or even as deeply as the *Gregaripus* tracks, also makes a mass stampede unlikely, because this gives us no reason to suspect that a carnivore was present when *Gregaripus* came through.

The plot of pace length vs. pace angle (fig. 10) shows that the *Gregaripus* locomotor pattern is strikingly different from that of the carnosaurs and *Grallator*(?). The pace length of *Gregaripus* is nearly constant, regardless of pace angle. As it accelerated, the legs tended to pull under the body and the track width narrowed. As it slowed, the legs tended to sprawl to the side and the track width widened. In both cases the pace remained nearly unchanged as the stride lengthened or shortened. To some degree, relative stride length increased as the animal sped up and pulled its legs in under its body (fig. 11). However this effect is much more pronounced in *Grallator*(?) and the anchisauripids than in *Gregaripus*. In its relative stride length, *Gregaripus* is quite similar to *Grallator*(?), but because of its smaller size, its striding rate is significantly higher (fig. 12), regardless of pace angle. This high striding rate allowed *Gregaripus* to attain speeds higher than any of the other dinosaurs studied here except *Grallator*(?), despite the fact that it was the smallest species recognized (fig. 13). Even so, the carnivore *Anchisauripus* could move as fast and with less exertion in the lower speed ranges of *Gregaripus* (fig. 13). This suggests that *Gregaripus* could outrun an anchisauripid over short distances, but over a long distance it would

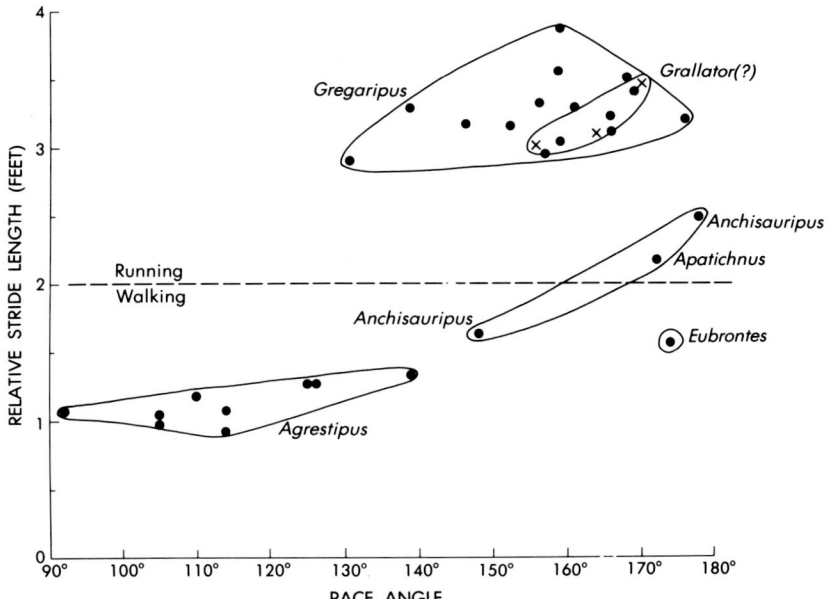

FIG. 11. Graph comparing the relative stride length of trackmakers from the Culpeper Crushed Stone Quarry with their pace angles. The boundary between walking and running is equivalent to a relative stride length of 2.0. In all species represented by more than one individual, pace angle tends to increase as stride length increases. This relationship is much more pronounced in carnivores (*Grallator*[?] and the anchisauripids) than in the herbivores (*Agrestipus* and *Gregaripus*). *Grallator*(?) values are shown by X's, all others by dots.

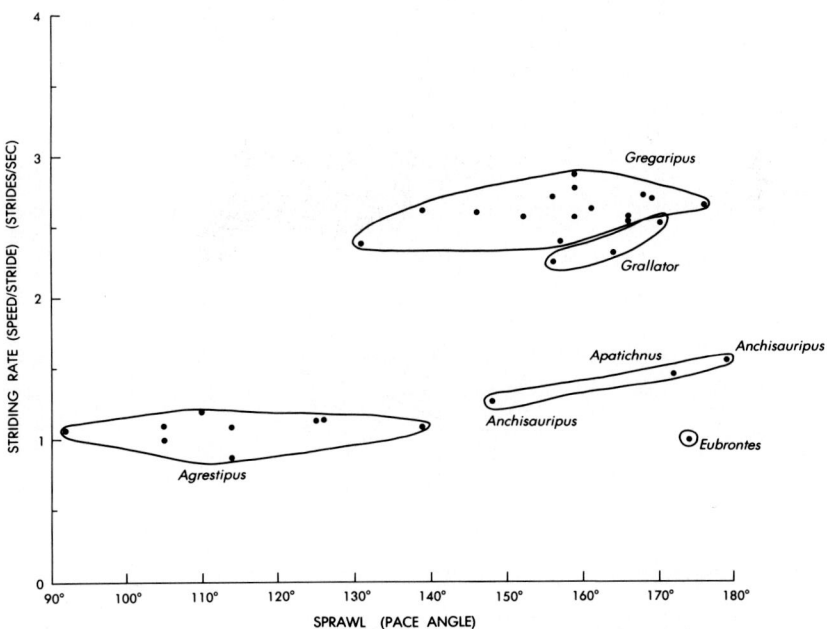

FIG. 12. Graph comparing the striding rate of trackmakers from the Culpeper Crushed Stone Quarry with their pace angles. In *Agrestipus* and *Gregaripus,* pace angle appears to be unrelated to the stepping rate.

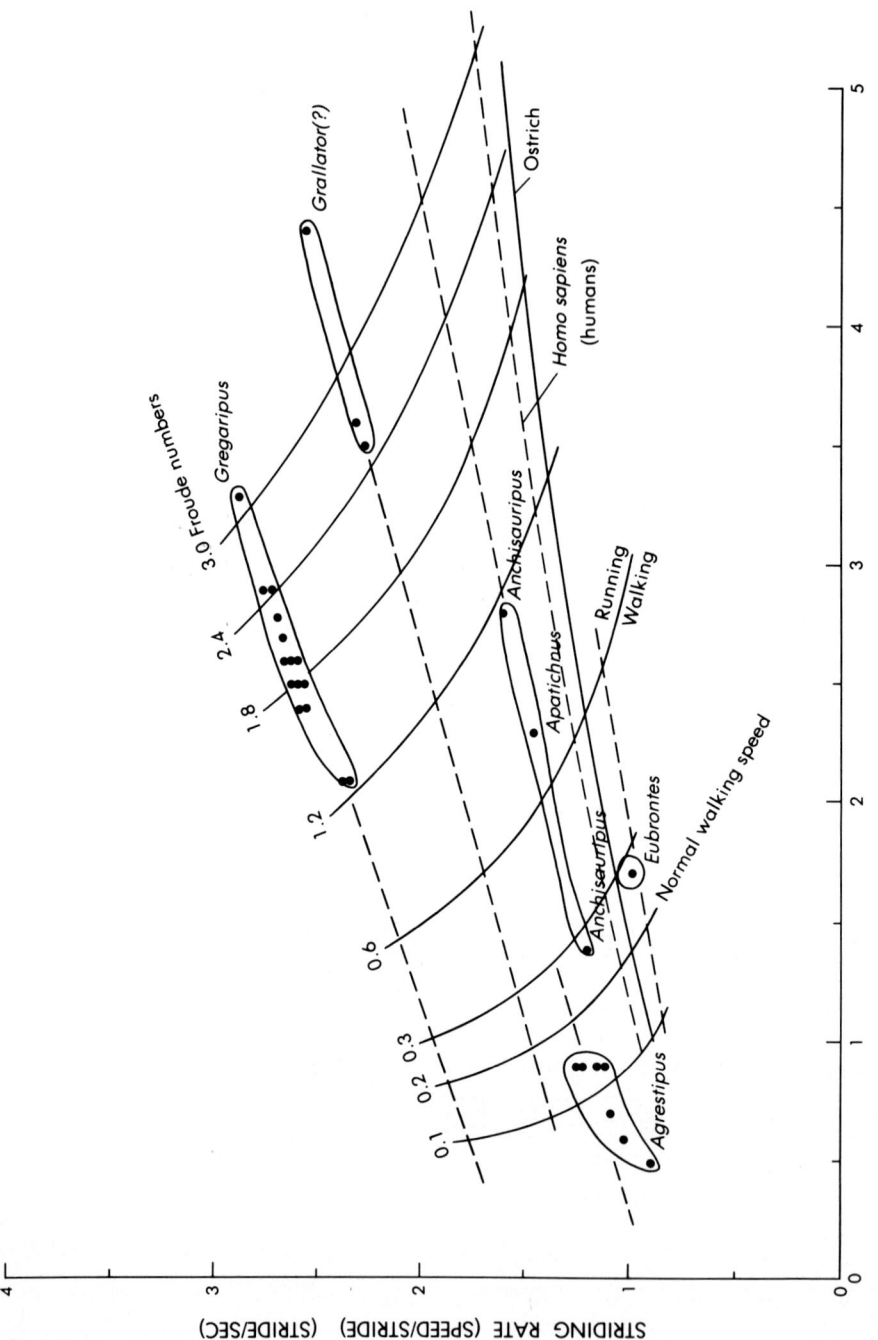

FIG. 13. Graph comparing the striding rate of trackmakers from the Culpeper Crushed Stone Quarry with their estimated speeds. The high stepping rate of *Gregaripus* reflects its small size (1.2 m long), which meant that it had to step more rapidly than the other forms to attain high speed. Conversely, *Eubrontes* only attained the speed it did by its large size (3.9 m long), which allowed it to cover long stretches of ground in relatively few steps. *Agrestipus, Anchisauripus, Apatichnus,* and *Grallator*(?) were all of similar length (0.5–0.7 m). A Froude number of 0.6 roughly marks the point at which animals cease walking and start running (Alexander, 1976). A Froude number of 0.2, the value for a human walking 3 miles per hour, is considered to represent normal walking speed. Only *Agrestipus*, which was the first animal to leave tracks on the bedding plane, appears to have moved slower than normal, possibly in response to poor footing on a very soft substrate. Froude number fields are plotted roughly on the basis of values derived from the studied trackways. These values were determined by using the formula F = v²/gh, where F is the Froude number, v is estimated velocity, g is acceleration due to gravity, and h is the estimated hip height (≃4 times the footprint length).

tire more quickly and might fall prey to one of these carnivores. These conclusions about the speed of *Gregaripus* indicate that it also was advanced in bipedal locomotion but in a quite different way from *Grallator*(?) and the carnosaurs. Because it was apparently capable of attaining considerable speed, despite its highly variable pace angle, the variable pace angle does not seem to have affected its running ability adversely.

Yet another pattern of locomotion is seen in *Agrestipus hottoni*. Its pattern of pace-angle rhythm is comparable in amplitude and irregularity with that of *Gregaripus* (fig. 8), though the pace-angle values are much lower. The pattern of the pace/pace angle field is similar in shape (fig. 10), but because both pace and pace angle are smaller in *Agrestipus*, the fields of the two genera do not overlap. In other respects, *Agrestipus* and *Gregaripus* were quite dissimilar (figs. 11–15). Assuming that *Agrestipus* had approximately the proportions of a bipedal dinosaur, the animal probably stood 0.6 m (2 ft) at the hips and would have been about 2.1 m (7 ft) long. Velocity values for this species are computed to have ranged from about 0.5 to 0.9 m/sec (1.1 to 2.1 mph). These ranges of speed are lower than the lowest values recorded by Alexander (1976) but do approach the speeds of sauropods. The Froude number for an animal should be about 0.2, when the animal is walking normally. Thus, with a relative stride length of 0.8 to 1.3 and Froude numbers of 0.04 to 0.14, *Agrestipus* must have been moving unusually ponderously. *Agrestipus* had a consistently low relative stride length relative to all the other forms except *Eubrontes*, which was walking in a similarly ponderous manner (fig. 14). Because of its short stride and relatively small size, *Agrestipus* was by far the slowest of all these animals (fig. 15). *Eubrontes*, in contrast, because of its much larger size, could cover ground at twice the rate of *Agrestipus*, even though its relative stride length suggests that it was only slightly more efficient proportionately in covering ground than was *Agrestipus*.

The major conclusions concerning the speeds and locomotor rates computed for all these forms are summarized in Table 1, and estimates for the normal speed, normal stride rate, and normal relative stride lengths are estimated from figures 13 and 15. When these data are normalized, all these forms should have had a relative stride length of about 1.4, the stride rate being inversely proportional to the hip height (and size) of the animals (as in pendulums), and the strolling speeds (which range from 0.8 to 1.5 m/sec) all being directly proportional to the animals' sizes. These trends are implicit in the formulas used.

Trackway Trends Through Time

Figure 16 is a graphic display of the average directions of travel for each species relative to the implied temporal succession of trackways discussed previously. The direction of travel can be appraised as an arithmetic mean either of all directions (360°) or of only half an arc (180°). For example, if animals went west (270°) and east (90°) along a trail, the average (180°)

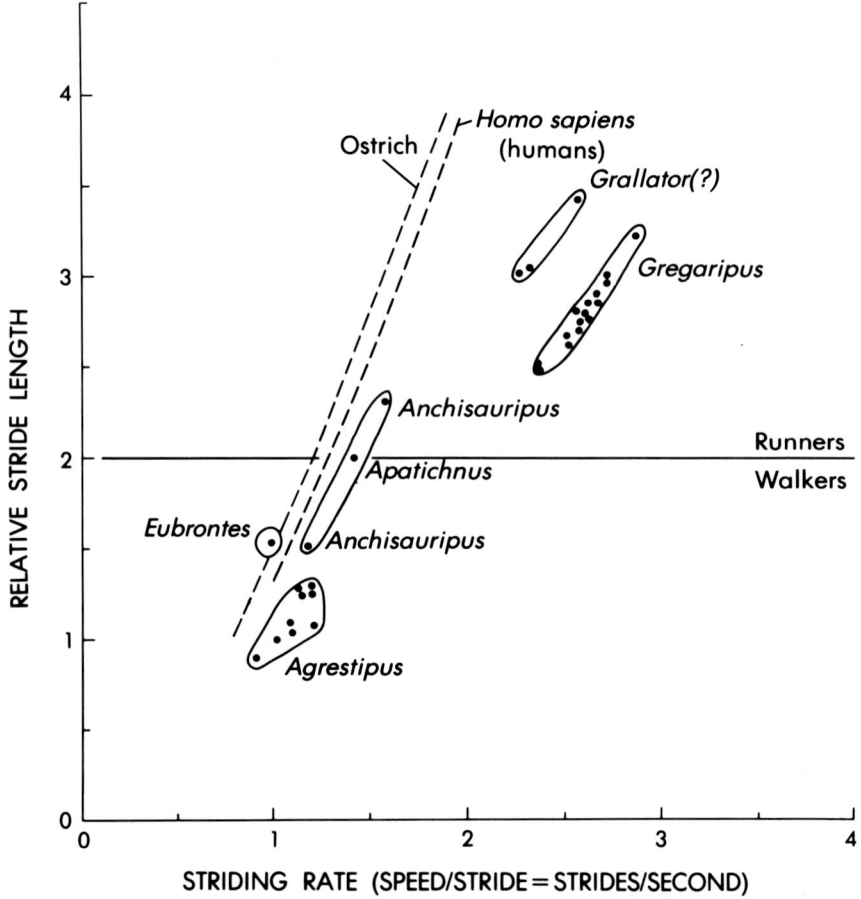

FIG. 14. Graph comparing the relative stride length of the trackmakers from the Culpeper Crushed Stone Quarry with their striding rate. The relationship for each species is nearly linear, and the slopes are quite similar. The position of the linear relationship for each species along the y-axis (stepping rate) is determined by the animal's absolute size, so that the species are smallest to right (*Gregaripus*) and largest to the left (*Eubrontes*).

is meaningless in assessing the true directions of movement, whereas a half-arc would portray the true trend of travel. To allow for this, both means were plotted. For the single trackway of *Apatichnus minor*, both values are, by definition, the same. They are also the same for *Agrestipus hottoni* (which has six directionally discernible trackways) as well as for *Anchisauripus parallelus* (three trackways), and are fairly close for *Gregaripus* (which has 15 directionally discernible trackways). The values are widely divergent for *Grallator*(?) and *Eubrontes*, however.

These patterns indicate that by either set of assumptions, all the *Agrestipus* and *Anchisauripus* were moving in the same general direction, though at widely different times, judging from the variations in track depth and clarity.

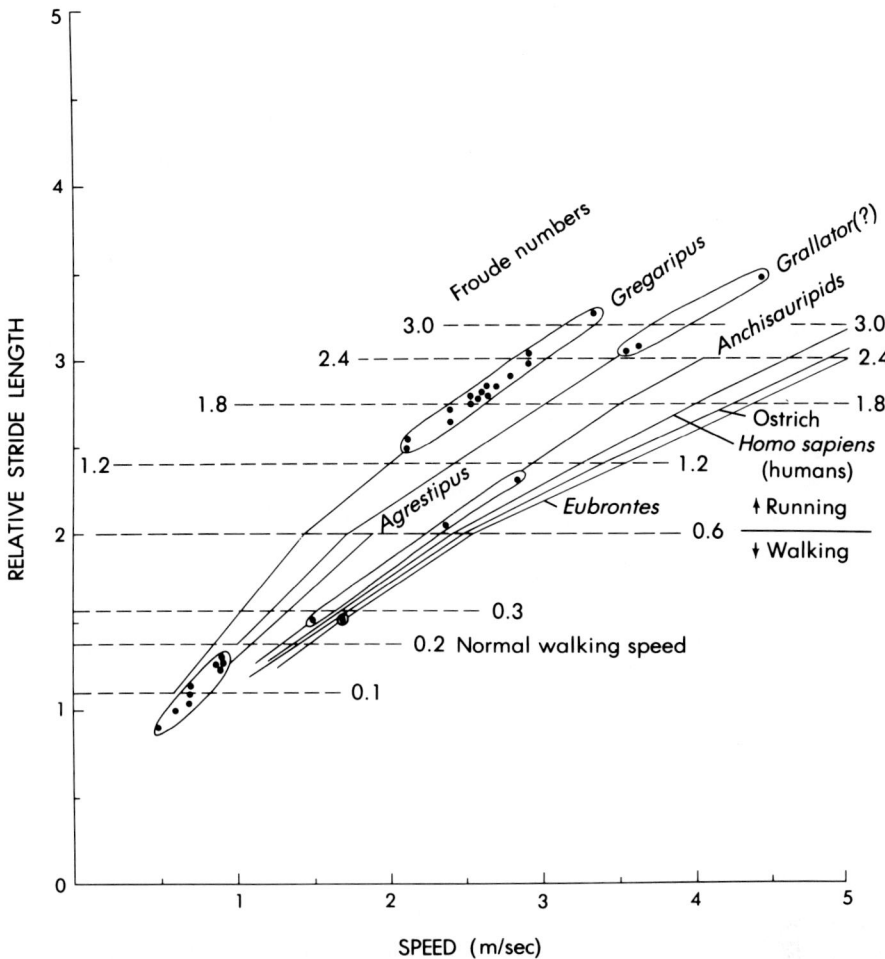

FIG. 15. Graph comparing the relative stride length of the trackmakers from the Culpeper Crushed Stone Quarry with their speeds (velocities) and Froude numbers. The correlation of Froude numbers with their corresponding relative stride lengths allows estimation of speeds not represented by actual trackways for each animal by v = Fgh (v = velocity, F = Froude number, h = hip height, g = acceleration due to gravity). Normal walking speed is defined as F = 0.2, the value for a human moving 3 mph. By this definition, *Agrestipus* shows an anomalously low relative stride length and speed, possibly because of poor footing, whereas *Gregaripus* and *Grallator*(?) were running quite rapidly for their size. Froude numbers were correlated to observed relative stride length by computing $F = v^2/gh$ for each trackway, and the relative stride length for each trackway was estimated by dividing stride length by four times the foot length.

Most of the *Gregaripus,* likewise, had a single direction of motion, although one was going in almost exactly the opposite direction, showing that both directions were being traveled. The compass divergence in the two plots on *Eubrontes* merely represents a single animal moving northwest on an east-oriented display, whereas the direction computed (163°) on only three widely divergent trackways of *Grallator*(?) is so close to random (180°) that

TABLE 1. Comparison of the Locomotor

Name	Suspected affinity	h Estimated hip height (meters)	l Estimated length (meters)	λ/h Relative stride length	F Estimated Froude number	V Estimated speed (meters/second)
Agrestipus	Sauropod	0.60	2.1	0.8–1.3	0.04–0.14	0.5–0.9
Anchisauripus	Carnosaur?	0.75	2.6	1.5–2.3	0.3–1.2	1.4–2.8
Apatichnus	Carnosaur?	0.75	2.6	2.0	0.7	2.3
Eubrontes	Carnosaur	1.10	3.9	1.5	0.3	1.7
Grallator (?)	Coelurosaur	0.50	1.8	3.0–3.5	2.5–3.9	3.5–4.4
Gregaripus	Ornithischian	0.35	1.2	2.5–3.3	1.3–3.1	2.1–3.3

it suggests no preferred direction at all. The net result of this plot by species indicates that the direction of travel by all these animals shifted systematically from SSE (154°) for the early arrivals to ESE (115°) for the later arrivals (dashed line on fig. 16), except for *Eubrontes* which was moving in an opposite but comparable sense. This shift suggests that this mudflat may have bordered a lake or river to the north, which was beginning to

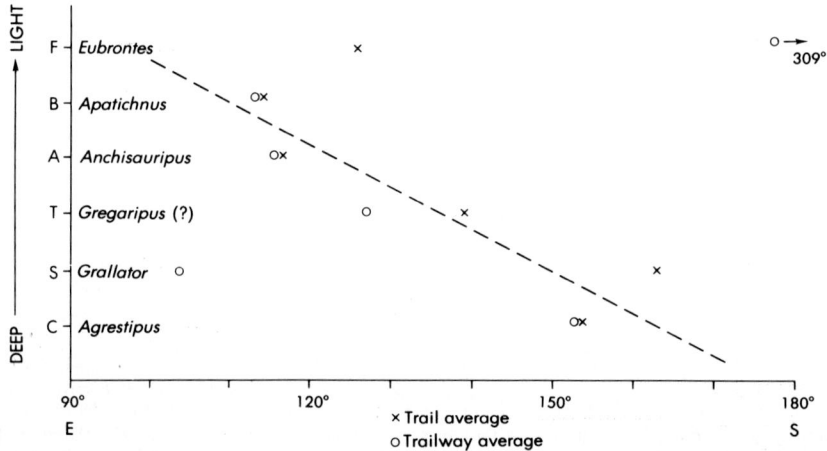

FIG. 16. Graph showing average direction of travel for each species found in the Culpeper Crushed Stone Quarry. "x"s mark directions computed using a 180° arc; "o"s are directions computed from a 360° arc. Vertical axis represents the relative order of each species' passage. "o" values assume that each direction of travel is independent from the direction taken by every other member of the same species; "x" values assume that individuals moving in opposite directions define a common pathway being travelled in both directions. The dashed diagonal line shows a systematic shift in the average direction of travel of each species, from south-southeasterly for early arrivals to easterly for late arrivals, except for *Eubrontes* which was moving opposite to this trend but in a comparable sense (x value). This shift in direction strongly suggests that the movement of these animals across this Triassic mudflat was not random.

Characteristics of the Species in This Study

Stride rate (stride/sec)	Travel mode	Estimated normal walking speed ($u^2/gh = 0.2$) (m/sec)	Estimated normal stride rate ($u^2/gh = 0.2$) (stride/sec)	Estimated normal relative stride length ($u^2/gh = 0.2$)
0.9–1.3	slow walking	1.1	1.3	1.4
1.2–1.6	trotting & walking	1.2	1.2	1.4
1.5	trotting?	1.2	1.2	1.4
1.0	walking	1.5	0.9	1.4
2.3–2.6	running	1.0	1.5	1.4
2.4–2.9	running	0.8	1.9	1.4

dry up. As the edge of the water receded northward, it permitted (or forced) the animals to follow a more easterly route while traveling around it. The well-structured order of appearance and direction of travel strengthen the idea that each species had a preferred substrate consistency, which tended to keep them ecologically separated.

Evidence for Gregariousness in Gregaripus

Although strongly directional trends are evident in trackways of most species, the variation around the mean is so wide that there is no indication of group travel. For *Gregaripus*, however, three closely related trends are evident (fig. 17). Each trend represents individuals that were in close proximity to each other (GE2, GE1, GE4, GE12, GE6, and GE7; GE10, GE8, GE9, and GE11; and GE13, GE14 and GE15 in fig. 3). These trackways are almost equally impressed on the bedding surface, a strong indication that these animals were traveling together in small groups that may or may not have constituted subunits of a large flock[1]. If the trackways of the largest group are arranged so that they are seen head-on and spaced apart in proportion to their actual spacing on the bedding plane (fig. 18), they show a systematic variation in average pace lengths. This pattern suggests that larger (long pace) individuals were traveling on either side of smaller (short pace) ones, possibly males guarding or herding females or else adults guarding or herding juveniles.

Evidence of flocking has been presented for the Early Jurassic dinosaur *Eubrontes* (Ostrom, 1971) and for Cretaceous hadrosaurs (Dodson, 1971; Currie and Sarjeant, 1979; Balsley, 1980; Lockley and others, 1983). Evidence of structured flocking has been presented for some Early Cretaceous sauropods (Bird, 1944). Therefore, the *Gregaripus* trackway patterns are

---

[1] "Flock" is used here because it is an archosaurian term, whereas the term "herd" is applied to mammals, which are synapsids.

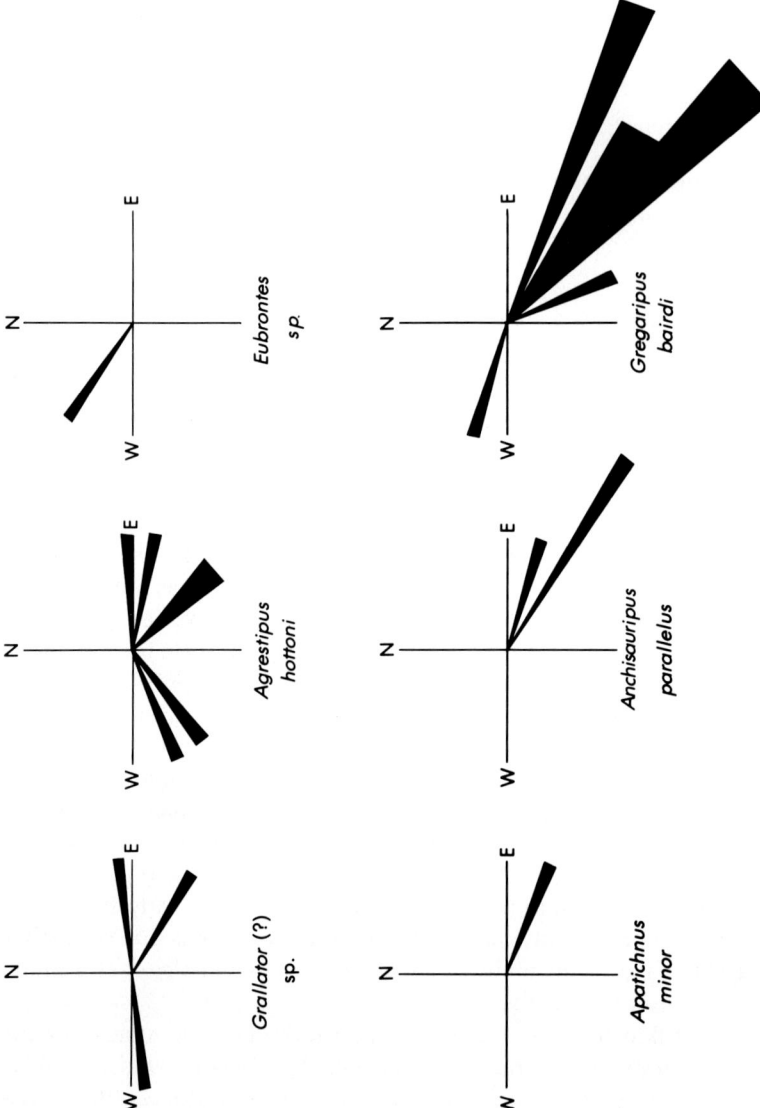

Fig. 17. Rose diagram showing directional trends of trackways (where known) for each species found in the Culpeper Crushed Stone Quarry. Length of the bars in *Agrestipus*, *Eubrontes*, and *Apatichnus* denotes single individuals.

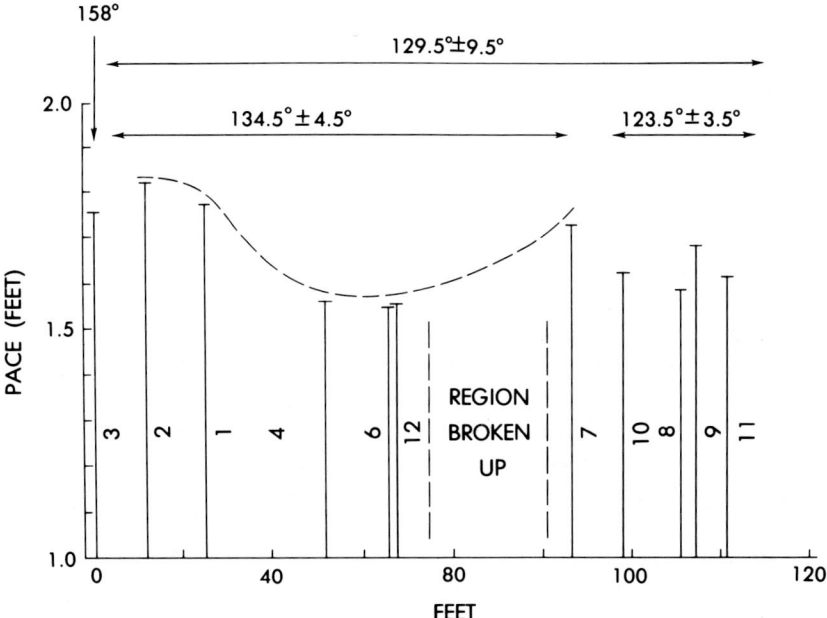

FIG. 18. Graph showing the relative spacing of the *Gregaripus* trails on the northern side of the Culpeper Crushed Stone Quarry (as measured along their intersection with, or projection to, trackway AG2/AG7) plotted against the average length of their pace. The top arrowed line shows the average directions of travel of the trackmakers (except for GE3) taken as a whole, plus or minus the *total* divergence from that trend of each individual. The lower arrowed lines show the same results if these are considered to represent two groups or subgroups. Dashed line shows the systematic distribution of different average pace lengths, suggesting that this group or subgroup was traveling in a structured pattern. A map view of this distribution can be seen in fig. 3.

comparable to the patterns shown in other groups of dinosaurs. *Gregaripus* is exceptional, however, because this is by far the oldest evidence reported of structured flocking among dinosaurs.

CONCLUSIONS

1) Trackways assigned to *Grallator*(?) and *Anchisauripus*, assumed in the past to have been made by carnivorous animals, are considered to have been correctly interpreted ecologically on the basis of their relative scarcity and efficient locomotor capabilities. *Apatichnus minor*, considered an ornithischian herbivore by Lull (1953), was probably a carnivore because of the similarity of its locomotor system to that of *Anchisauripus*. The animal which made the trackway here designated as *Eubrontes* sp. is also considered to be a carnivore, though at least some animals which made trackways elsewhere called *Eubrontes* probably were not carnivores.

2) Trackways assigned to *Agrestipus* and *Gregaripus* are considered to have been made by herbivores, on the basis of their relatively great abundance and their blunt-toed feet.

3) From evidence indicating sustained rapid locomotion, sustained functional endothermy can be inferred for *Grallator*(?) and *Gregaripus*. The apparent overabundance of large carnivore trackways (*Anchisauripus*, *Apatichnus*, *Eubrontes*), indicated by a high predator/prey ratio, could have occurred because there was a collecting bias in favor of the large carnivore trackways or because the carnivores were lurking in a favorite hunting spot. But alternatively, it is also possible that the large carnivores were poikilothermic in contrast to the small carnivores and herbivores. Endothermy in *Agrestipus* cannot be evaluated, but its relatively ponderous gait gives no hint of such a level of sophistication.

4) When *Grallator*(?) and *Gregaripus* passed through the area exposed in the quarry, they were running; *Apatichnus* and one of the *Anchisauripus* were trotting; all others were walking.

5) The locomotor systems of the carnosaurs (*Anchisauripus*, *Apatichnus*, *Eubrontes*), sauropods (*Agrestipus*), coelurosaurs (*Grallator*[?]) and ornithischians (*Gregaripus*) recognized at this site were all distinctly different by Norian time, judging from study of their respective trackways. This indicates quite different lifestyles for each of these groups, even at this early time in the history of dinosaurs.

6) *Grallator*(?), which was surprisingly agile and sure-footed on a soft muddy surface, may have had feathers. This raises the possibility that all coelurosaurs may have been feathered, and may therefore be more properly classified as birds than as dinosaurs. Although probably not capable of flight, this animal seemed to develop sufficient aerodynamic lift to run easily at high speed and to maneuver quickly over very soft

36

ground. Because of its consistently fast running gait, it probably needed to be nearly or entirely endothermic.

7) *Gregaripus* represents one of the oldest known ornithischian dinosaurs. It has almost the same dimensions as *Heterodontosaurus tucki* and probably was similar in appearance. This animal generally traveled in flocks, some of which may have been structured so that smaller individuals were kept near the middle of the group for protection. Because the trackways indicate that all 15 of the *Gregaripus* at the Culpeper locality had been running, these animals also were probably functional endotherms.

8) *Agrestipus*, which has a brachycheirotheroid track, may be ancestral to the sauropods. In later sauropods, the emphasis on digits I–III in the pes would be a logical result of the relative lengths of the toes seen in this form. The strong development of digit III in the theropod and plateosaur pes, and the concomitant relative atrophy of I (and V) makes it difficult to search for the ancestry of *Agrestipus* among those groups, though *Agrestipus* may be derived from a generalized prosauropod stock. Although *Agrestipus* appears to have been bipedal as it came across the bedding plane studied here, it is likely that it was not obligately bipedal judging from its exceptionally low pace angle.

9) Although most major lineages of dinosaurs appear to be represented in this Norian footprint assemblage, all of these animals are miniatures of their later Mesozoic descendants. Relative sizes within the community are in accord with later patterns except for *Agrestipus*, which is a rather small animal for a sauropod. The consistency of their relative sizes implies that the dinosaurs were a well-established and highly sophisticated community even as early as Norian time.

10) Most of the trackways described here were made by animals which were walking (11 individuals with relative stride length less than 2.0) or running (18 individuals with relative stride length greater than 2.9). Only two trackways represent individuals moving at a slow run or trot. This observation is in accord with Thulborn's (1984) observation that dinosaurs appear to have avoided a trotting gait except as an unavoidable transition between walking and running.

11) The speeds of travel estimated here are mostly within the range of speeds estimated by Alexander (1976), Kool (1981), Thulborn (1981), and Farlow (1981). However, Farlow (1981) and Thulborn (1984) have documented a few Cretaceous trackways which appear to have been made by dinosaurs moving at speeds in the range of 20–40 km/hr. Moreover, Thulborn (1982), on anatomical grounds, has shown that speed capabilities up to 35–40 km/hr are to be expected. The fact that most Jurassic and Cretaceous dinosaurs, however, moved at speeds comparable to the speeds of the Norian dinosaurs studied here is curious. Perhaps dinosaurs, for the most part, did not improve their speed capabilities as they evolved toward larger sizes. This sort of pattern

has been noted in horses. Eocene forms, which were much smaller than their modern descendants, probably were very nearly as fast as horses are now (Simpson, 1961, p. 251–267). Alternatively, there may be simply a limit to the rate of stable progression which can be attained on soft substrates. There is no *a priori* reason to expect that the speeds attained by dinosaurs on soft ground reflect their maximum speed capabilities on firm ground.

Acknowledgments

The author would like to thank Laurel Bybell, Kenneth Carpenter, Stewart Fagin, Albert Froelich, David Govoni, Nicholas Hotton III, Marguerite Kingston, Vincent LaPiana, Daniel Milton, Terri Purdy, Joseph Smoot and Byron Stone for the extensive help and advice which they all contributed toward the conduct of this research and preparation of this paper. Special thanks also go to the management and employees of the Culpeper Crushed Stone Quarry, without whose help and cooperation this study would not have been possible.

REFERENCES CITED

Alexander, R. McN., 1976, Estimates of speeds of dinosaurs: Nature (London), v. 261, p. 129–130.

Baird, Donald, 1954, *Cheirotherium lulli,* a pseudosuchian reptile from New Jersey: Mus. Comp. Zool. Harvard Coll., Bull., v. III, no. 4, p. 165–192.

——1957, Triassic reptile footprint faunules from Milford, New Jersey: Mus. Comp. Zool. Harvard Coll., Bull., v. 117, no. 5, p. 449–520.

——1980, A prosauropod dinosaur trackway from the Navajo Sandstone (Lower Jurassic) of Arizona, *in* Aspects of Vertebrate History (L. L. Jacobs, ed.), Museum of Northern Arizona Press, Flagstaff, p. 219–230.

Bakker, R. T., 1975, Experimental and fossil evidence for the evolution of tetrapod bioenergetics, *in* Perspectives in biophysical ecology (eds., D. Gates & R. Schmerl): New York, Springer-Verlag, p. 365–399.

Balsley, J. K., 1980, Cretaceous wave dominated delta systems, Book Cliffs, eastern central Utah: American Association of Petroleum Geologists field seminar guidebook, p. 163.

Bird, R. T., 1944, Did *Brontosaurus* ever walk on land?: Nat. Hist., v. 53, p. 61–67.

Bock, Wilhelm, 1952, Triassic reptilian tracks and trends of locomotive evolution, with remarks on correlation: Journal of Paleontology, v. 26, no. 3, p. 395–433.

Carozzi, A. V., 1964, Complex ooids from Triassic lake deposit, Virginia: Amer. Jour. Sci., v. 262, no. 2, p. 231–241.

Charig, A. J., Attridge, J., and Crompton, A. W., 1965, On the origin of the sauropods and the classification of the saurischia: Linn. Soc. London Proc., v. 176, no. 2, p. 197–221.

Chatterjee, Sankar, 1985, *Postosuchus,* a new thecodontian reptile from the Triassic of Texas and the origin of tyrannosaurs: Philosophical Transactions of the Royal Society of London, vol. B309, p. 395–460.

Colbert, E. H., 1965, A phytosaur from North Bergen, New Jersey: Amer. Museum Novitates, no. 2230, p. 1–25.

Cornet, Bruce, 1977, The palynostratigraphy and age of the Newark Supergroup: University Park, Pa., Pennsylvania State Univ., Ph.D. thesis, p. 1–505.

Cornet, Bruce, Traverse, A., and McDonald, N., 1973, Fossil spores, pollen and fish indicate Early Jurassic age for part of the Newark Group: Science, v. 182, p. 1243–1247.

Currie, P. J., and Sargeant, W. A. S., 1979, Lower Cretaceous dinosaur footprints from the Peace River Canyon, British Columbia, Canada: Palaeogeography, Palaeoclimatology, Palaeoecology, v. 28, p. 103–115.

Dodson, Peter, 1971, Sedimentology and taphonomy of the Oldman Formation (Campanian), Dinosaur Provincial Park, Alberta (Canada): Palaeogeography, Palaeoclimatology, Palaeoecology, v. 10, p. 21–74.

Emmons, Ebenezer, 1857, American geology, part 6: Albany, N.Y., Sprague and Co., p. 1–152.

Farlow, J. O., 1981, Estimates of dinosaur speeds from a new trackway site in Texas: Nature, v. 294, p. 747–748.

Haubold, Hartmut, 1971, Ichnia amphibiorum et reptiliorum fossilium, *in* Handbuch der Palaeoherpetologie (O. Kuhn, ed.): Stuttgart, Gustav Fischer Verlag, part 18, p. 1–124.

Huene, Frederik von, 1908, Die Dinosaurier der Europaischen Triasformation mit Berucksichtigung der Aussereuropaischen Vorkommnisse: Geologische und Palaeontologische Abhandlungen, v. 1 (supplement), 419 p.

—— 1926, Vollstandige Osteologie eines Plateosauriden aus dem Schwabischen Keuper: Geologische und Palaeontologische Abhandlungen (n.s.), v. 15, no. 2, p. 139–180.

Hitchcock, Edward, 1858, Ichnology of New England. A report on the sandstone of the Connecticut valley, especially its fossil footmarks: Boston, Wm. White, p. 1–220.

Jain, S. L., Kutty, T. S., Roychowdhury, T., and Chatterjee, S., 1977, Some characteristics of *Barapasaurus tagorei,* a sauropod dinosaur from the Lower Jurassic of Deccan, India: IV International Gondwana Symposium, 1977, Calcutta, India, p. 204–216.

Kingston, M. J., LaPiana, V. J., Purdy, T. L. and Weems, R. E., 1976, Footprint evidence for flocking behavior in Late Triassic bipedal reptiles: Geol. Soc. Amer., Abstracts with Programs, v. 8, no. 2, p. 19.

Kool, Richard, 1981, The walking speed of dinosaurs from the Peace River Canyon, British Columbia, Canada: Canadian Journal of Earth Sciences, v. 18, no. 4, p. 823–825.

Lee, K. Y., 1977, Triassic stratigraphy in the northern part of the Culpeper basin, Virginia and Maryland: U.S. Geol. Survey Bull. 1422-C, p. 1–17.

Lindholm, R. C., 1979, Geologic history and stratigraphy of the Triassic-Jurassic Culpeper basin, Virginia: Geol. Soc. America Bull., v. 90, no. 11, p. I995–I997, II1702–II1736.

Lockley, M. G., Young, B. H., and Carpenter, Kenneth, 1983, Hadrosaur locomotion and herding behavior: evidence from footprints in the Mesaverde Formation, Grand Mesa coal field, Colorado: Mountain Geologist, v. 20, n. 1, p. 5–14.

Luca, Albert P. S., Crompton, A. W., and Charig, Alan J., 1976, A complete skeleton of the late Triassic ornithischian *Heterodontosaurus tucki:* Nature (London), v. 264, p. 324–328.

Lull, R. S., 1953, Triassic life of the Connecticut River valley (revised): Connecticut Geol. Nat. Hist. Survey Bull. 81, p. 1–331.

Olsen, P. E., Remington, C. L., Cornet, B., and Thomson, K. S., 1978, Cyclic change in Late Triassic lacustrine communities: Science, v. 201, p. 729–733.

Ostrom, J. H., 1971, Were some dinosaurs gregarious?: Palaeogeography, Palaeoclimatology, Palaeoecology, v. 11, p . 287–301.

Pannell, N. K., 1985, Dinosaur footprints at Oak Hill, Virginia: Masters Thesis, The George Washington University, Washington, D.C., 30 p.

Roberts, J. K., 1928, The geology of the Virginia Triassic: Virginia Geological Survey Bulletin, v. 29, p. 1–205.

Sargent, W. A. S., 1975, Fossil tracks and impressions of vertebrates, *in* The study of trace fossils (R. W. Frey, ed.): New York, Springer-Verlag, p. 283–324.

Shaler, N. S., and Woodworth, J. B., 1899, Geology of the Richmond basin, Virginia: U.S. Geol. Survey, 19th Ann. Rept., v. 2, p. 393–515.

Simpson, G. G., 1961, *Horses:* Garden City, N.Y., Doubleday and Company, 323 p.

Thulborn, R. A., 1981, Estimated speed of a giant bipedal dinosaur: Nature, v. 292, p. 273–274.

—— 1982, Speeds and gaits of dinosaurs: Palaeogeography, Palaeoclimatology, Palaeoecology, v. 38, p. 227–256.

—— 1984, Preferred gaits of bipedal dinosaurs: Alcheringa, v. 8, p. 243–252.

Thulborn, R. A., and Wade, Mary, 1979, Dinosaur stampede in the Cretaceous of Queensland: Lethaia, v. 12, p. 275–279.

Wanner, H. E., 1926, Some additional faunal remains from the Trias of York County, Pennsylvania: Acad. Nat. Sci. Philadelphia Proc., v. 78, p. 1–14.

Weems, R. E., 1979, A large parasuchian (phytosaur) from the upper Triassic portion of the Culpeper basin of Virginia (USA): Biol. Soc. Washington Proc., v. 92, no. 4, p. 682–688.

—— 1980, An unusual, newly discovered archosaur from the Upper Triassic of Virginia, U.S.A.: Amer. Phil. Soc. Trans., v. 70, no. 7, p. 1–53.

Young, Chung-Chien, 1941, A complete osteology of *Lufengosaurus heunei* Young (gen. et sp. nov.) from Lufeng, Yunnan, China: Palaeontologia Sinica (n.s.), Series C, no. 7, 53 p.

Young, R. S., and R. S. Edmundson, 1954, Oolitic limestone in the Triassic of Virginia: Journal of Sedimentary Petrology, v. 24, no. 4, p. 275–279.

APPENDIX

Measurements on trackways from the Culpeper Crushed Stone Quarry. Left column is distance between prints (as preserved), right column is compass bearing between successive prints taken by Brunton compass. Missing tracks are marked with an "X." Trackway patterns are visually summarized in fig. 19.

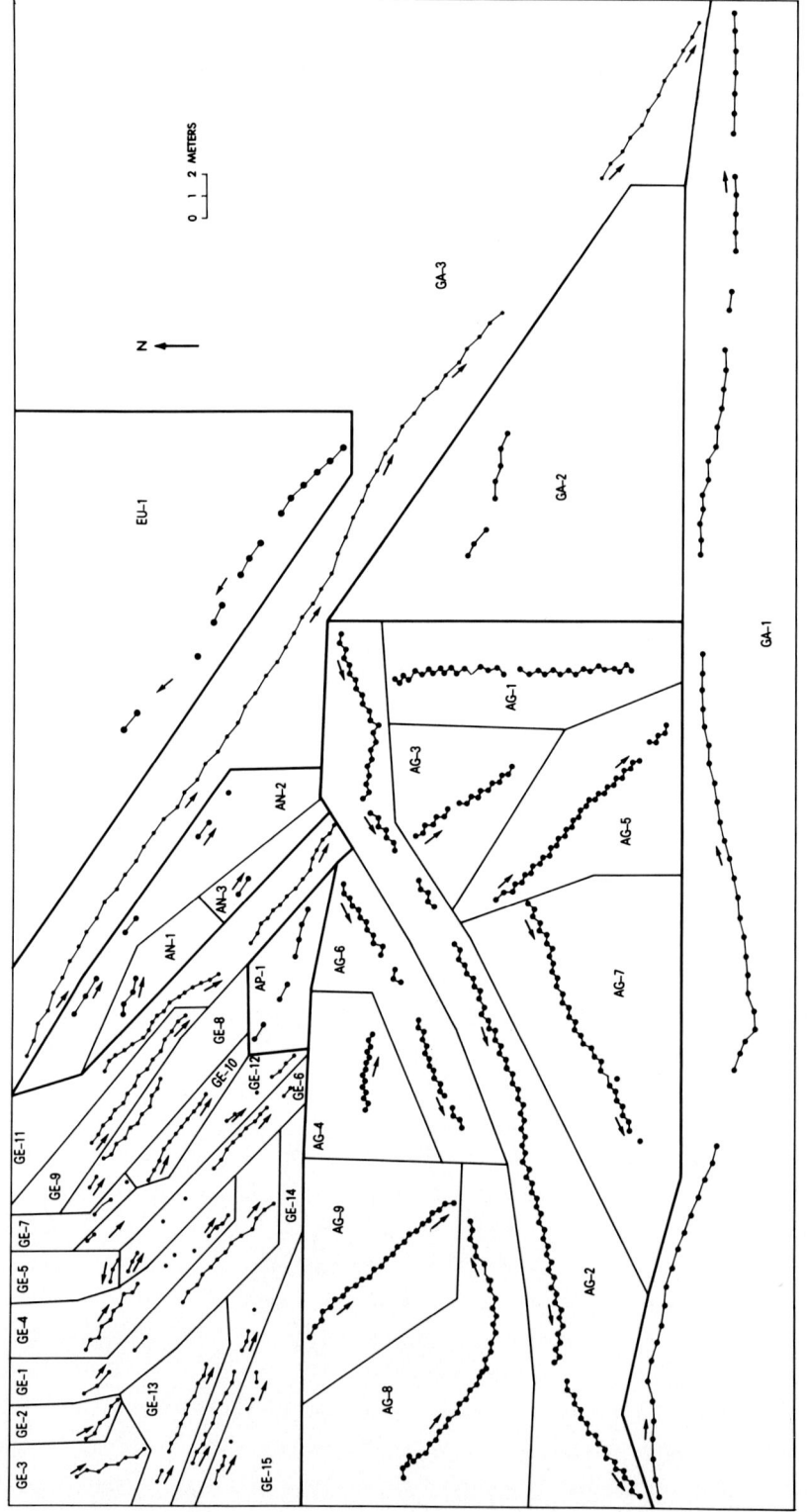

FIG. 19. Visual display of trackways recorded in Appendix 1. Trackway numbers match the numbers used in fig. 3.

TRACKWAY NO.:   <u>Agrestipus</u> 1

Pace: 0.45   Stride: 0.65   Pace angle: 92°

Trackway trend: 180°-360° (direction of travel unknown)

Pace (meters)	Compass bearing	Pace angle

```
                                                            "End"
                                                            South

0.42    ┌ 222
        o--------- 108                            0.33   │  230
0.51    │ 150                                            o---------73
        o--------- 120                            0.57   │  123
0.42    │ 210                                            o---------84
        o--------- 100                            0.33   │  219
0.60    │ 130                                            o--------121
        o---------- 91                            0.51   │  160
0.45    │ 219                                            o--------118
        o--------- 95                             0.39   │  222
0.54    │ 134                                            o---------92
        o--------- 80                             0.57   │  134
0.42    │ 234                                            o--------104
        o---------- 81                            0.39   │  210
0.54    │ 135                                            o--------105
        o--------- 67                             0.57   │  135
0.33    │ 248                                            o---------96
        o--------- 65                             0.42   │  219
0.54    │ 133                                            o--------103
        o--------- 80                             0.51   │  142
0.36    │ 233                                            o--------101
        o--------- 75                             0.45   │  221
0.51    │ 128                                            o--------101
        o--------- 86                             0.51   │  142
0.45    │ 222                                            o--------100
        o--------- 75                             0.39   │  222
0.66    │ 117                                            o--------105
        o--------- 70                             0.51   │  147
0.39    │ 227                                            o--------119
        o--------- 66                             0.57   │  208
0.42    │ 113                                            o
        o--------- 71                                    │
0.27    │ 222                                            o
        │                                                │
        o                                        0.84    X    162
                                                         │
                                                         X
      "Start"                                            │
      North                                              o
```

34 tracks   X track missing   Speed: 0.7 m/s (=2.5 km/hr=1.5 mi/hr)

44

TRACKWAY NO.:   <u>Agrestipus</u> 2 (probably=<u>Agrestipus</u> 7)

Pace: 0.42   Stride: 0.75   Pace angle: 125°

Trackway trend: 234°

Pace (meters)	Compass bearing	Pace angle		Pace (meters)	Compass bearing		(cont.)		
0.43	↑ 281			0.41	↑ 258		0.45	216	
	o------------------ 114				o			o----------124	
0.48	215				X		0.42	272	
	o------------------ 117				X			o----------99	
0.48	278			1.71	228		0.42	191	
	o------------------ 112				X			o----------108	
0.47	210						0.41	263	
	o------------------ 114				X			o----------107	
0.47	276						0.42	190	
	o------------------ 105			0.41	264			o----------113	
0.46	201				o--------115		0.43	257	
	●------------------ 111			0.45	199			o----------126	
0.50	270				o--------105		0.45	203	
	●------------------ 119			0.41	274			o----------107	
0.43	209				o--------98		0.47	276	
	●------------------ 120			0.47	192			o----------119	
0.45	269				o--------93		0.42	215	
	●------------------ 122			0.45	279			o	
0.45	211				o--------102			X	
	o------------------ 118			0.44	201			X	
0.45	273				o		2.03	240	
	o------------------ 118			0.79	X 240			X	
0.44	211				o			X	
	o------------------ 114			0.43	280			o	
0.45	277				o--------105		0.41	208	
	o------------------ 121			0.45	205			o----------112	
0.44	218				o--------123		0.47	276	
	o------------------ 110			0.45	262			o----------98	
0.43	288				o--------130		0.43	194	
	o------------------ 106			0.47	212			o----------116	
0.42	214				o--------111				
	o------------------ 118								
0.47	276								
	o------------------ 100								
0.47	196								
	o								

Start
NE

111 tracks   X track missing   ● cast prints   speed: 0.9 m/s (=3.2 km/hr=1.9 mi/hr)

TRACKWAY NO.: <u>Agrestipus</u> 2 (cont.)

Pace (meters)	Compass bearing	Pace angle
0.42	216	
		116
0.45	280	
		112
0.43	212	
		117
0.43	275	
		109
0.42	204	
		121
0.43	263	
		118
0.41	201	
		127
0.48	254	
		123
0.43	197	
		113
0.42	264	
		114
0.47	198	
		108
0.43	270	
		113
0.43	203	
		123
0.45	260	
		94
0.33	174	
		88
0.39	266	
		109
0.46	195	
		96
0.45	279	
		117

(cont.)

Pace (meters)	Compass bearing	Pace angle
0.46	195	
		102
0.45	273	
		107
0.45	200	
		111
0.43	269	
		117
0.47	206	
		121
0.41	265	
		119
0.46	204	
		113
0.43	271	
		114
0.42	205	
		120
0.43	265	
		110
0.43	195	
		116
0.43	259	
		122
0.42	201	
		109
0.43	272	
		109
0.46	201	
		100
0.45	281	
		116
0.43	217	
		116
0.45	281	
		115

(cont.)

Pace (meters)	Compass bearing	Pace angle
0.42	278	
		106
0.47	204	
		109
0.43	275	
		110
0.47	205	
		115
0.42	270	
1.0	X 231	
0.41	287	
		101
0.43	206	
		100
0.41	286	
		115
0.42	221	
		118
0.47	283	
		107
0.43	210	
		123
0.43	267	
		109
0.47	196	
		116
0.46	260	
		122
0.41	202	
		117
0.45	265	
		110

TRACKWAY NO.: <u>Agrestipus</u> 2 (cont.)

Pace (meters)	Compass bearing	Pace angle

```
              End
              SW
              ↑
              o
0.46          |      213
              o------------------ 114
0.37          |      279
              o--------------- 111
0.38          |      210
              o------------------ 121
0.42          |      269
              o------------------ 107
0.42          |      196
              o--------------- 109
0.43          |      267
              o------------------ 117
0.45          |      204
              o--------------- 104
0.43          |      280
              o--------------- 104
0.46          |      204
              o------------------ 120
0.43          |      264
              o--------------- 117
0.43          |      201
              o--------------- 104
0.41          |      277
              o------------------ 121
0.41          |      218
              o================= 120
```

(cont.)

TRACKWAY NO.: <u>Agrestipus</u> 3

Pace: 0.40   Stride: 0.63   Pace angle: 105°

Trackway trend: 144°

Pace (meters)	Compass bearing	Pace angle

```
        End
        SE

         ↑
         o
0.42     |     109
         o------------------ 116
0.41     |     173
         o------------------ 101
0.37     |     94
         o------------------ 97
0.41     |     179
         o------------------ 110
0.37     |     109
         o------------------ 108
0.41     |     181
         o------------------ 98
0.37     |     99
         o------------------ 86
0.41     |     193
         o------------------ 95
0.39     |     108
         o------------------ 103
0.41     |     185
         o
         |
0.61     X     154
         |
         o
0.41     |     122
         o------------------ 119
0.41     |     183
         o------------------ 112
0.38     |     115
         o------------------ 115
0.43     |     180
         o------------------ 92
0.41     |     92
         o------------------ 113
0.41     |     159
         o
        Start
        NW
```

18 tracks   X track missing   speed: 0.7 m/s (=2.4 km/hr=1.4 mi/hr.)

TRACKWAY NO.: Agrestipus 4

Pace: 0.33   Stride: 0.55   Pace angle: 114°

Trackway trend: 89°

Pace (meters)	Compass bearing	Pace angle

```
          End
          East
           ↑
0.33       |     121
           o------------------ 118
0.33       |      59
           o------------------ 124
0.33       |     115
           o------------------ 114
0.30       |      49
           o------------------ 104
0.33       |     125
           o------------------ 141
0.30       |      86
           o------------------ 146
0.33       |     120
           o------------------ 109
0.33       |      49
           o------------------ 107
0.33       |     122
           o------------------ 98
0.30       |      40
           o------------------ 105
0.33       |     115
           o------------------ 104
0.36       |      39
           o------------------ 101
0.33       |     118
           o

          Start
          West
```

14 tracks   speed: 0.5 m/s (=1.9 km/hr=1.1 mi/hr)

49

TRACKWAY NO.: Agrestipus 5

Pace: 0.37   Stride: 0.59   Pace angle: 105°

Trackway trend: 135°

Pace (meters)	Compass bearing	Pace angle
0.33	358	106
0.36	284	100
0.36	004	91
0.36	275	105
0.39	350	104
0.36	274	90
0.36	004	89
0.42	273	95
0.36	358	103
0.36	281	116
0.33	345	
X		
0.96	298	
X		
0.36	357	93
0.33	270	103
0.42	347	108
0.39	275	

Start NW

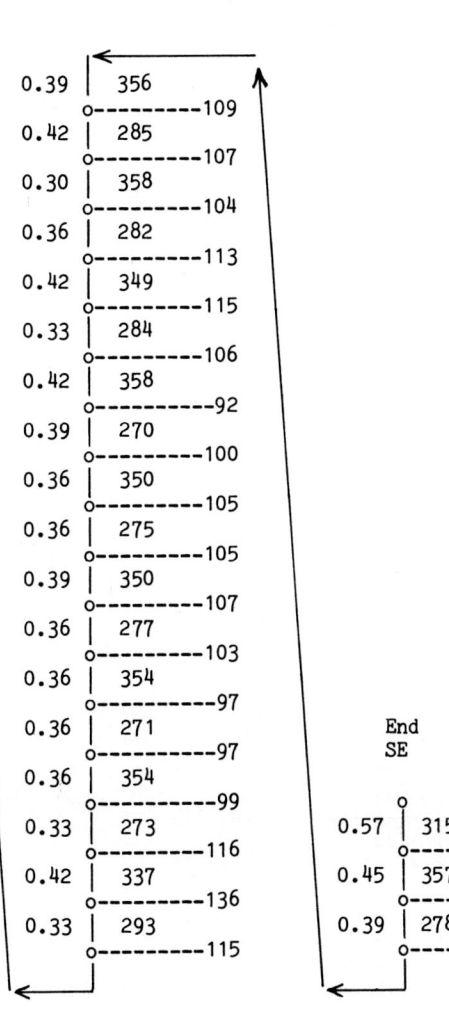

0.39	356	109
0.42	285	107
0.30	358	104
0.36	282	113
0.42	349	115
0.33	284	106
0.42	358	92
0.39	270	100
0.36	350	105
0.36	275	105
0.39	350	107
0.36	277	103
0.36	354	97
0.36	271	97
0.36	354	99
0.33	273	116
0.42	337	136
0.33	293	115

End SE

0.57	315	138
0.45	357	101
0.39	278	102

38 tracks   X track missing   speed: 0.6 m/s (=2.1 km/hr=1.3 mi/hr)

TRACKWAY NO.: Agrestipus 6

Pace: 0.39   Stride: 0.63   Pace angle: 114°

Trackway trend: 245°

Pace (meters)	Compass bearing	Pace angle

End
SW
↑

				End SW ↑

```
Pace          Compass        Pace                    End
(meters)      bearing        angle                   SW
                                                      ↑
              ↑ ─────────────────────→       0.42  | o  260
 2.04         X   242                                 o────────119
              |                              0.39  |    199
              X                                       o────────118
              |                              0.42  |    261
              o                                       o
 0.33         |   199                        0.60  X    241
              o─────────────── 104                 |
 0.42         |   275                              o
              o                              0.39  |    208
              |                                       o────────104
              X                              0.45  |    284
 0.96         |   224                                 o─────────97
              X                              0.42  |    201
              |                                       o────────108
              o                              0.42  |    273
 0.42         |   278                                 o────────116
              o─────────────── 88            0.33  |    207
 0.51         |   186                                 o────────109
              o─────────────── 92            0.36  |    278
 0.42         |   274                                 o────────121
              o─────────────── 108           0.36  |    219
 0.42         |   202                                 o────────111
              o─────────────── 119           0.39  |    288
 0.42         |   263                                 o────────116
              o─────────────── 120           0.39  |    224
 0.36         |   203                                 o────────122
              o─────────────── 119           0.42  |    282
 0.39         |   264                                 o────────110
              o─────────────── 118           0.36  |    212
 0.39         |   202                                 o────────119
              o─────────────── 130           0.39  |    273
 0.39         |   252                              o
              o─────────────── 136                 |
 0.39         |   208                              X
              o─────────────── 123                 |
 0.42         |   265                              X
              o                                    |
              |                                    X

              Start                                ↓   X
              NE                                       ────
```

32 tracks   X track missing   speed: 0.7 m/s (=2.4 km/hr=1.4 mi/hr)

TRACKWAY NO.: _Agrestipus_ 7 (probably = _Agrestipus_ 2)

Pace: 0.45   Stride: 0.74   Pace angle: 110°

Trackway trend: 238°

Pace (meters)	Compass bearing	Pace angle			Pace			
0.42	278						End SW	
		104						
0.42	202						o	
		108						
0.48	274							
		106			0.66	X	215	
0.42	200						o	
		105						
0.45	275				0.45		260	
		120					130	
0.45	215				0.45		210	
		115					120	
0.45	280				0.39		270	
		115				●	115	
0.45	215				0.42		205	
		106					125	
0.42	289				0.45		260	
		115					127	
0.42	224				0.45		207	
		107					132	
0.42	297				0.45		255	
		91					128	
0.42	208				0.45		203	
		110					107	
0.45	278				0.48		276	
		96					99	
0.45	194				0.45		195	
		93					124	
0.42	281				0.45		251	
		111					118	
0.48	212				0.45		189	
		106					117	
0.42	286				0.45		252	
		106					138	
0.42	212				0.45		210	
							112	

Start
NE

34 tracks   X track missing   speed: 0.9 m/s(=3.1 km/hr=1.8 mi/hr)
            ● track cast

TRACKWAY NO.: <u>Agrestipus</u> 8

Pace: 0.43   Stride: 0.76   Pace angle: 126°

Trackway trend starts at 135°, ends at 69° (av. 103°)

Pace (meters)	Compass bearing	Pace angle			End NE
	↑ ———————————→				↑ o
?0.21	↑ 120?		0.42		65
	o—————————— 156				o——————140
0.42	96		0.45		105
	o—————————— 137				o——————117
0.42	139		0.48		42
	o—————————— 145				o——————133
0.42	104		0.45		89
	o—————————— 141				o——————123
0.45	143		0.48		32
	o—————————— 128				o——————130
0.42	91		0.45		82
	o—————————— 126				o——————136
0.54	145		0.39		38
	o—————————— 132				o——————129
0.36	97		0.45		89
	o—————————— 131				o——————138
0.42	146		0.51		47
	o—————————— 137				o——————122
0.42	103		0.45		105
	o—————————— 135				o——————124
0.42	148		0.42		49
	o—————————— 138				o——————121
0.42	106		0.45		108
	o—————————— 124				o——————131
0.39	162		0.42		59
	o—————————— 148				o——————126
0.36	130		0.48		113
	o—————————— 88				o——————146
0.27	222		0.45		79
	o—————————— 58				o——————136
0.42	94		0.48		123
	o—————————— 111				o——————127
0.30	163		0.45		70
	o—————————— 94				o——————107
0.54	77		0.24		143
	o				o——————157
	Start NW				

37 tracks   speed: 0.9 m/s(=3.2 km/hr=1.9 mi/hr)

TRACKWAY NO.: <u>Agrestipus 9</u>

Pace: 0.42   Stride: 0.78   Pace angle: 139°

Trackway trend: 130°

Pace (meters)	Compass bearing	Pace angle
0.42	152	
		130
0.42	102	
		135
0.42	147	
		140
0.42	107	
		145
0.42	142	
		143
0.42	105	
		130
0.42	155	
		131
0.42	106	
		125
0.45	161	
		144
0.39	125	
		146
0.45	159	
		131
0.39	110	
		134
0.42	156	
		143
0.45	119	
		151
0.42	148	
		123
0.42	91	
		136
0.45	134	
		161
0.45	115	

Start
NW

End
SE

0.45	158	
		139
0.39	117	
		148
0.42	149	
		133
0.39	102	
		141
0.39	141	
		152
0.42	113	
		141

25 tracks   speed: 0.9 m/s(=3.4 km/hr=2.0 mi/hr)

TRACKWAY NO.: <u>Anchisauripus</u> 1 (Possibly continuous with <u>Anchisauripus</u> 2)

Pace: 0.6   Stride: 1.15 Pace angle: 148°

Trackway trend: 106°

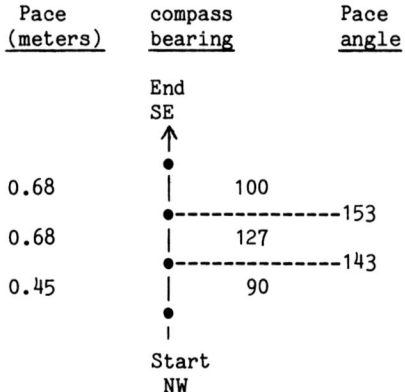

Pace (meters)	compass bearing	Pace angle
	End SE ↑ ●	
0.68	| 100	
	●--------------153	
0.68	| 127	
	●--------------143	
0.45	| 90	
	● ⎮ Start NW	

4 tracks   ● track cast   speed: 1.4 m/s(5.0 km/hr=3.1 mi/hr)

TRACKWAY NO.: _Anchisauripus_ 2 (possibly continuous with _Anchisauripus_ 1)

Pace: 0.87   Stride: 1.74 Pace angle: 179°

Trackway trend: 123°

```
   Pace            Compass       Pace
 (meters)          bearing       angle

                   End
                   SE
                   ↑
                   o
                   |
   1.65            X    120
                   |
                   o
   0.86            |    125
                   o
                   |
                   X
                   |
                   X
   4.74            |    125
                   X
                   |
                   X
                   |
                   o
   0.81            |    126
                   o
                   |
                   X
   2.61            |    121
                   X
                   |
                   o
   0.88            |    122
                   o---------------179
   0.91            |    121
                   o
                   |
                   Start
                   NW
```

8 tracks   X track missing   speed: 2.8 m/s(=10.1 km/hr=6.3 mi/hr)

56

Pace: 0.85   Stride: ?   Pace angle: ?

Trackway trend: 122°

Pace (meters)	Compass bearing	Pace angle

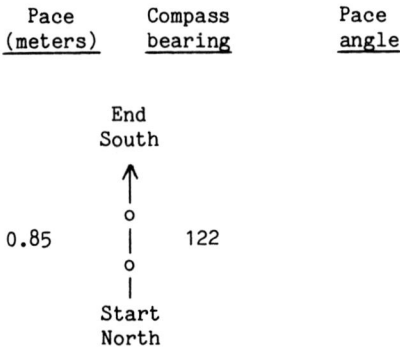

```
          End
          South
           ↑
           o
0.85       |      122
           o
           |
          Start
          North
```

2 tracks

TRACKWAY NO.: <u>Apatichnus</u> 1

Pace: 0.77   Stride: 1.54 Pace angle: 172°

Trackway trend: 114°

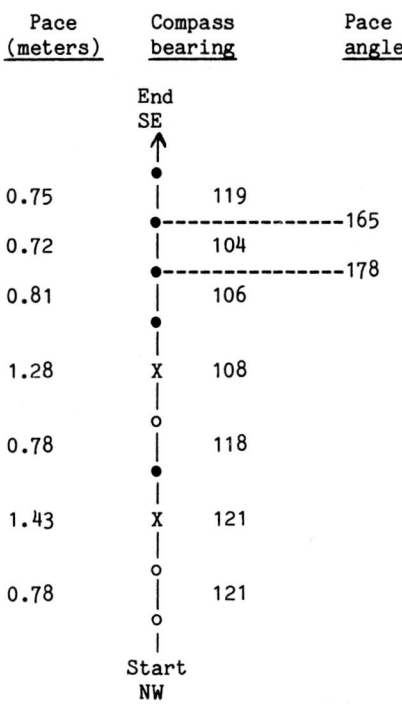

Pace (meters)	Compass bearing	Pace angle
	End SE ↑ ●	
0.75	| 119	
	●---------------165	
0.72	| 104	
	●---------------178	
0.81	| 106	
	● |	
1.28	X 108	
	| o	
0.78	| 118	
	● |	
1.43	X 121	
	| o	
0.78	| 121	
	o |	
	Start NW	

8 tracks   X track missing speed: 2.3 m/s(=8.3 km/hr=5.2 mi/hr)
● track cast

58

Pace: 0.85   Stride: 1.70   Pace angle: 174°
Trackway trend: 309°

Pace (meters)	Compass bearing	Pace angle
	End NW ↑	
	○	
0.99	│ 312	
	●	
	│	
	X	
	│	
3.84	X 320	
	│	
	X	
	│	
	○	
	│	
1.80	X 296	
	│	
	○	
0.84	│ 297	
	○	
	│	
1.71	X 302	
	│	
	○	
about 0.84+	│ 297	
	○----------------------171	
0.85	│ 306	
	○	
	│	
1.69	X 305	
	│	
	○	
0.86	│ 303	
	●--------------------167	
0.84	│ 316	
	○--------------------180	
0.85	│ 316	
	○------------------176	
0.80	│ 320	
	○------------------174	
0.81	│ 314	
	○	
	Start SE	

                                              + Smudged

14 tracks   X track missing   ● track cast   speed: 1.7 m/s(=6.2 km/hr=3.7 mi/hr)

TRACKWAY NO.:   Grallator 1

Pace: 0.87   Stride: 1.73   Pace angle: 170°

Trackway trend: 84° (264°)

Column 1

Pace (meters)	Compass bearing	Pace angle
0.93	265	
1.80 X	272	
0.87	275	
1.80 X	275	
0.90	265	176
0.87	269	180
0.90	269	177
0.87	266	
1.86 X	268	
0.93	268	179
0.90	269	179
0.96	270	178
0.96	268	177
0.96	265	175
0.87	270	

Start
East

Column 2

Pace (meters)	Compass bearing	Pace angle
0.84	270	
X		
X		
4.56 X	269	
X		
X		
0.72	268	173
0.72	261	159
0.69	282	160
0.69	262	152
0.72	290	156
0.87	264	147
0.78	297	151
0.90	268	168
0.87	280	172
0.87	272	177
0.87	275	170

Column 3 (cont.)

Pace (meters)	Compass bearing	Pace angle
0.87	241	
		157
0.87	264	175
0.87	259	180
0.87	259	169
0.93	248	166
0.90	262	170
0.90	252	177
0.90	255	174
0.87	249	173
0.90	256	175
0.90	261	178
0.84	259	176
0.90	255	167
0.84	268	167
0.84	255	170
0.81	265	180
0.84	265	175
0.81	270	180

71 tracks   X track missing   speed: 4.4 m/s(=15.8 km/hr=9.9 mi/hr)

Trackway no.:   Grallator 1 (cont.)

Pace (meters)	Compass bearing	Pace angle
0.96	276	175
0.87	271	166
0.87	285	180
0.90	285	173
0.84	292	175
0.84	287	173
0.87	294	174
0.87	288	172
0.87	296	169
0.87	285	
1.74	X 287	
1.74	X 279	
0.87	282	177
0.57	285	160
0.72	305	135
0.84	260	161

(cont.)

End
West

0.93	250	175
0.90	255	174
0.93	261	174
0.87	267	173
0.87	260	170
0.93	250	155
0.93	275	171
0.84	284	176
0.78	280	176

TRACKWAY NO.: <u>Grallator</u> 2

Pace: 0.77   Stride: 1.51   Pace angle: 156°

Trackway trend: 285°

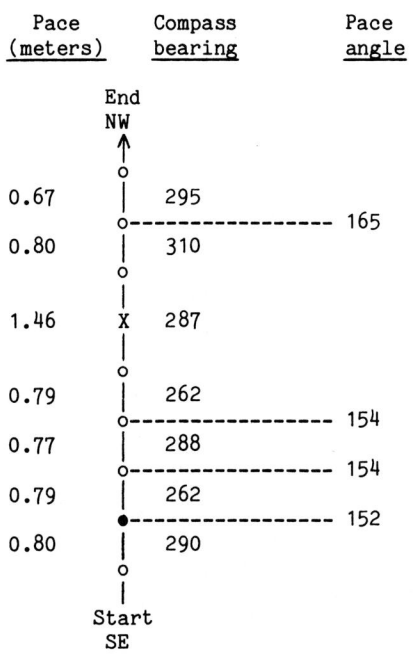

Pace (meters)	Compass bearing	Pace angle

```
              End
              NW
               ↑
               o
0.67           |    295
               o----------------- 165
0.80           |    310
               o
               |
1.46           X    287
               |
               o
0.79           |    262
               o----------------- 154
0.77           |    288
               o----------------- 154
0.79           |    262
               ●----------------- 152
0.80           |    290
               o
               |
              Start
              SE
```

8 tracks   X track missing   ● track cast   speed: 3.5 m/s(=12.6 km/hr=7.9 mi/hr)

62

TRACKWAY NO.: Grallator 3

Pace: 0.78   Stride: 1.54   Pace angle: 164°

Trackway trend: 121°

Pace (meters)	Compass bearing	Pace angle				(cont.)	
0.79	119		0.79	103			
		154			172	X	
0.81	145		0.81	111			
		165			170	X	
0.81	130		0.79	101			
		180			157	o	
0.79	130		0.76	124		0.76	134
		172			173		162
0.81	122		0.79	117		0.81	116
		168			167		165
0.81	130		0.81	130		0.76	131
		169			165		160
0.81	119		0.79	115		0.79	111
		165			174		162
0.76	134		0.76	121		0.79	129
		160			180		166
0.81	114		0.79	121		0.81	115
		165			171		162
0.76	129		0.81	130		0.79	133
		168			170		165
0.79	117		0.79	120		0.79	118
		173			168		170
0.76	124		0.81	132		0.76	128
		169			162		160
0.79	113		0.84	114		0.76	108
		171			162		165
0.81	122		0.84	132		0.79	123
		165			170		164
0.81	107		0.84	122		0.81	107
		159			161		170
0.81	128		0.84	141		0.81	117
		154			158		164
0.79	102		0.81	119		0.81	101
		166			159		168
0.84	116		0.81	140		0.81	113
					159		170

Start

63 tracks   X track missing   speed: 3.6 m/s(=13.0 km/hr=8.1 mi/hr)

TRACKWAY NO.: <u>Grallator</u> 3 (cont.)

Pace (meters)	Compass bearing	Pace angle

```
                    End
                     ↑
                     o
    0.76             |     105
                     o----------------- 173
    0.79             |     112
                     o----------------- 176
    0.79             |     108
                     o----------------- 174
    0.81             |     114
                     o----------------- 166
    0.76             |     100
                     o----------------- 165
    0.76             |     115
                     o----------------- 165
    0.76             |     100
                     o----------------- 156
    0.76             |     124
                     o----------------- 161
    0.76             |     105
                     o----------------- 160
    0.74             |     135
                     o----------------- 163
    0.76             |     118
                     o
                     |
                     X
                     |
                     X
                     |
                     X
                     |
                     X
                     |
    7.76             X     127
                     |
                     X
                     |
                     X
                     |
                     X
                     |
                  (cont.)
```

TRACKWAY NO.: <u>Gregaripus</u> 1

Pace: 0.54   Stride: 1.06   Pace angle: 159°
Trackway trend: 134°

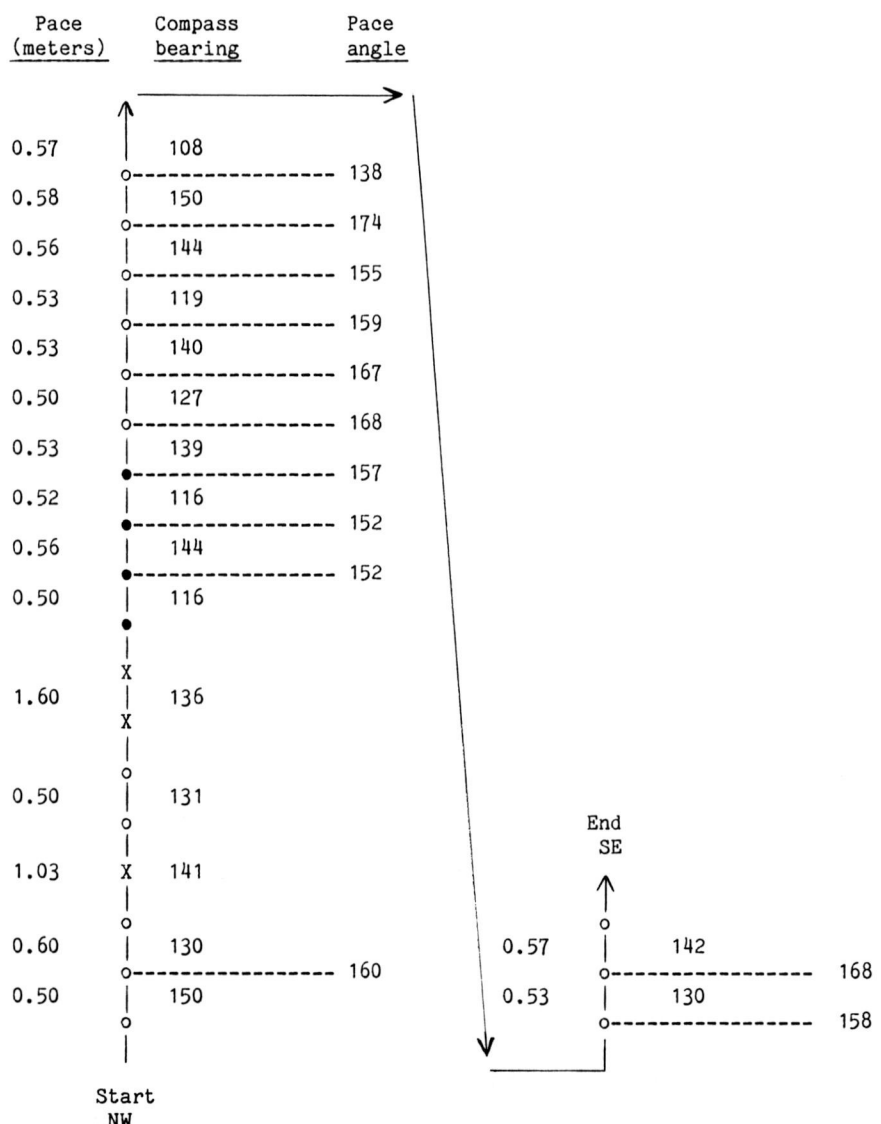

Pace (meters)	Compass bearing	Pace angle

18 tracks   X track missing   ● track cast   speed: 2.9 m/s(=10.4 kmhr=6.3 mi/hr)

TRACKWAY NO.: <u>Gregaripus</u> 2

Pace: 0.50   Stride: 0.97   Pace angle: 152°

Trackway trend: 130°


```
   Pace        Compass       Pace
 (meters)      bearing       angle

              End
              SE
               ↑
               o
  0.48         |    136
               o----------------- 155
  0.51         |    111
               o----------------- 151
  0.48         |    140
               o----------------- 155
  0.52         |    115
               o----------------- 145
  0.50         |    150
               o
               |
              Start
              NW
```


6 tracks   speed: 2.5 m/s(=9.0 km/hr=5.4 mi/hr)

TRACKWAY NO.: <u>Gregaripus</u> 3

Pace: 0.53   Stride: 0.99   Pace angle: 139°

Trackway trend: 158°

Pace (meters)	Compass bearing	Pace angle
	End SE ↑	
	o	
0.53	| 133	
	o---------------- 135	
0.56	| 178	
	o---------------- 138	
0.48	| 136	
	o---------------- 138	
0.53	| 178	
	o---------------- 143	
0.53	| 141	
	o---------------- 135	
0.58	| 186	
	o---------------- 147	
0.53	| 153	
	o	
	|	
	Start NW	

8 tracks   speed: 2.6 m/s(=9.4 km/hr=5.6 mi/hr)

TRACKWAY NO.: <u>Gregaripus</u> 4

Pace: 0.48   Stride: 0.87   Pace angle: 131°

Trackways trend: 137°

```
  Pace          Compass           Pace
(meters)        bearing           angle

                    ↑  ─────────────────→
                    o
                    |
 0.97               X     134
                    |
                    o
                    |
 0.95               X     169
                    |
                    o
 0.66               |     132
                    o
                    |
                    X
 1.57               |     147
                    X
                    |
                    o
 0.47               |     115
                    o──────────────── 126
 0.50               |     169
                    o──────────────── 111
 0.52               |     100
                    o──────────────── 138
 0.47               |     142
                    o──────────────── 137
 0.50               |     99                    End
                    o──────────────── 135       SE
 0.47               |     144                    ↑
                    o──────────────── 137        o
 0.48               |     101                     |
                    o──────────────── 117    0.47 |125
 0.48               |     164                     o─────────141
                    o──────────────── 124    0.39 |164
 0.47               |     108                     o─────────139
                    o                        0.50 |123
                    |
                  Start
                   NW
```

17 tracks   X track missing   speed: 2.1 m/s(=7.6 km/hr=4.5 mi/hr)

68

TRACKWAY NO.: <u>Gregaripus</u> 5

Pace: 0.58   Stride: 1.14   Pace angle: 159°

Trackway trend: 289°

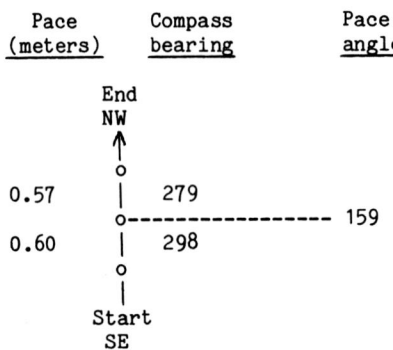

```
     Pace        Compass        Pace
   (meters)      bearing        angle

                   End
                   NW
                    ↑
                    o
    0.57            |    279
                    o------------------ 159
    0.60            |    298
                    o
                    |
                  Start
                    SE
```

3 tracks   speed: 3.3 m/s(=11.9 km/hr=7.1 mi/hr)

TRACKWAY NO.: <u>Gregaripus</u> 6

Pace: 0.47   Stride: 0.93   Pace angle: 166°

Trackway trend: 131°

Pace (meters)	Compass bearing	Pace angle

```
        Pace        Compass      Pace
      (meters)      bearing      angle

                      ↑    ─────────────────→
                  o─────────────── 169
        0.46      │      135
                  o─────────────── 176
        0.50      │      131
                  o─────────────── 166
        0.50      │      117
                  o─────────────── 165
        0.50      │      132
                  o
                  │
                  X
                  │
        1.98      X      132
                  │
                  X
                  │
                  o
                  │
        1.44      X      134
                  │
                  X
                  │                          End
                  o                          SE
                  │                           ↑
                  X                  0.46      o      128
                  │                           │
        1.93      X      130                  o
                  │                  0.93      X      144
                  X                           │
                  │                           o
                  o                  0.47      │      138
                  │                           o───────────169
                  o                  0.44      │      127
                  │                           o───────────161
                  o                  0.48      │      146
        0.47      │      124                  o───────────158
                  o─────────────── 165  0.46      │      124
        0.43      │      109
                Start
                 NW
```

16 tracks   X track missing   speed: 2.4 m/s(=8.6 km/hr=5.2 mi/hr)

TRACKWAY NO.: <u>Gregaripus</u> 7

Pace: 0.53   Stride: 1.05   Pace angle: 168°

Trackway trend: 135°-315° (direction of travel unknown)

```
    Pace        Compass      Pace
  (meters)      bearing      angle

                  |
   1.14           o    132
                  |
                  X
                  |
                  o
                  |
   0.50           |    144
                  o------------------- 168
   0.56           |    132
                  o
                  |
```

3 tracks   X track missing   speed: 2.9 m/s(=10.4 km/hr=6.3 mi/hr)

TRACKWAY NO.: Gregaripus 8

Pace: 0.50   Stride: 0.96   Pace angle: 146°

Trackway trend: 120°

Pace (meters)	Compass bearing	Pace angle
	End SE ↑	
	o	
0.43	| 98	
	o----------------- 159	
0.48	| 119	
	o----------------- 148	
0.46	| 87	
	o----------------- 140	
0.52	| 127	
	o----------------- 158	
0.51	| 105	
	o----------------- 139	
0.46	| 156	
	o----------------- 134	
0.51	| 110	
	o----------------- 142	
0.46	| 148	
	o----------------- 147	
0.55	| 115	
	o----------------- 150	
0.44	| 145	
	o	
	|	
0.85	X 115	
	|	
	o	
0.66	| 121	
	o	
	|	
	Start NW	

13 tracks   speed: 2.5 m/s(=9.0 km/hr=5.4 mi/hr)

TRACKWAY NO.: <u>Gregaripus</u> 9

Pace: 0.51   Stride: 1.00   Pace angle: 156°

Trackway trend: 126°

```
    Pace      Compass        Pace
  (meters)    bearing       angle

               End
               SE
                ↑
                o
   0.51         |    145
                o---------------- 154
   0.51         |    119
                o---------------- 170
   0.52         |    129
                o---------------- 155
   0.55         |    104
                o---------------- 150
   0.55         |    134
                o---------------- 161
   0.52         |    115
                o---------------- 153
   0.56         |    142
                o---------------- 148
   0.51         |    110
                o---------------- 156
   0.50         |    134
                o---------------- 154
   0.46         |    108
                o---------------- 144
   0.51         |    144
                o---------------- 155
   0.50         |    119
                o---------------- 161
   0.48         |    138
                o---------------- 155
   0.48         |    113
                o---------------- 164
   0.56         |    129
                o
                |
              Start
               NW
```

16 tracks   speed: 2.7 m/s(=9.7 km/hr=5.8 mi/hr)

TRACKWAY NO.: <u>Gregaripus</u> 10

Pace: 0.49   Stride: 0.98   Pace angle: 176°

Trackway trend: 122°

```
    Pace        Compass         Pace
  (meters)      bearing         angle

                 End
                 SE
                  ↑
                  o
   0.48          |    120
                  o----------------- 176
   0.45          |    124
                  o----------------- 177
   0.45          |    127
                  o----------------- 180
   0.48          |    127
                  o----------------- 177
   0.45          |    124
                  o----------------- 177
   0.48          |    127
                  o----------------- 180
   0.51          |    127
                  o----------------- 163
   0.54          |    110
                  o----------------- 173
   0.48          |    117
                  o----------------- 178
   0.57          |    115
                  o
                  |
                 Start
                 NW
```

11 tracks   speed: 2.6 m/s(=9.4 km/hr=5.6 mi/hr)

TRACKWAY NO.: <u>Gregaripus</u> 11

Pace: 0.48   Stride: 0.94   Pace angle: 159°

Trackway trend: 127°

Pace (meters)	Compass bearing	Pace angle		End SE	
				○	
0.48	135		0.45	116	
		170			154
0.45	145		0.45	142	
		180			158
0.48	145		0.48	120	
		171			178
0.45	136		0.51	118	
		179			165
0.36	135		0.54	103	
		175			153
0.54	140		0.57	130	
		155			175
0.54	105		0.48	125	
		155			165
0.39	140		0.54	140	
		157			165
0.45	117		0.51	125	
		175			160
0.39	112		0.42	145	
		172			155
0.45	119		0.42	120	
		152			160
0.60	147		0.51	140	
		171			150
0.51	138		0.45	110	
		173			166
0.54	145		0.54	124	
		172			
0.39	137		1.11 X	140	
		158			
0.51	115				
			0.99 X	134	
0.87 X	145				

Start
NW

34 tracks   X track missing   speed: 2.4 m/s (=8.6 km/hr=5.2 mi/hr)

TRACKWAY NO.: _Gregaripus_ 12

Pace: 0.49   Stride: 0.97   Pace angle: 166°

Trackway trend: 139°

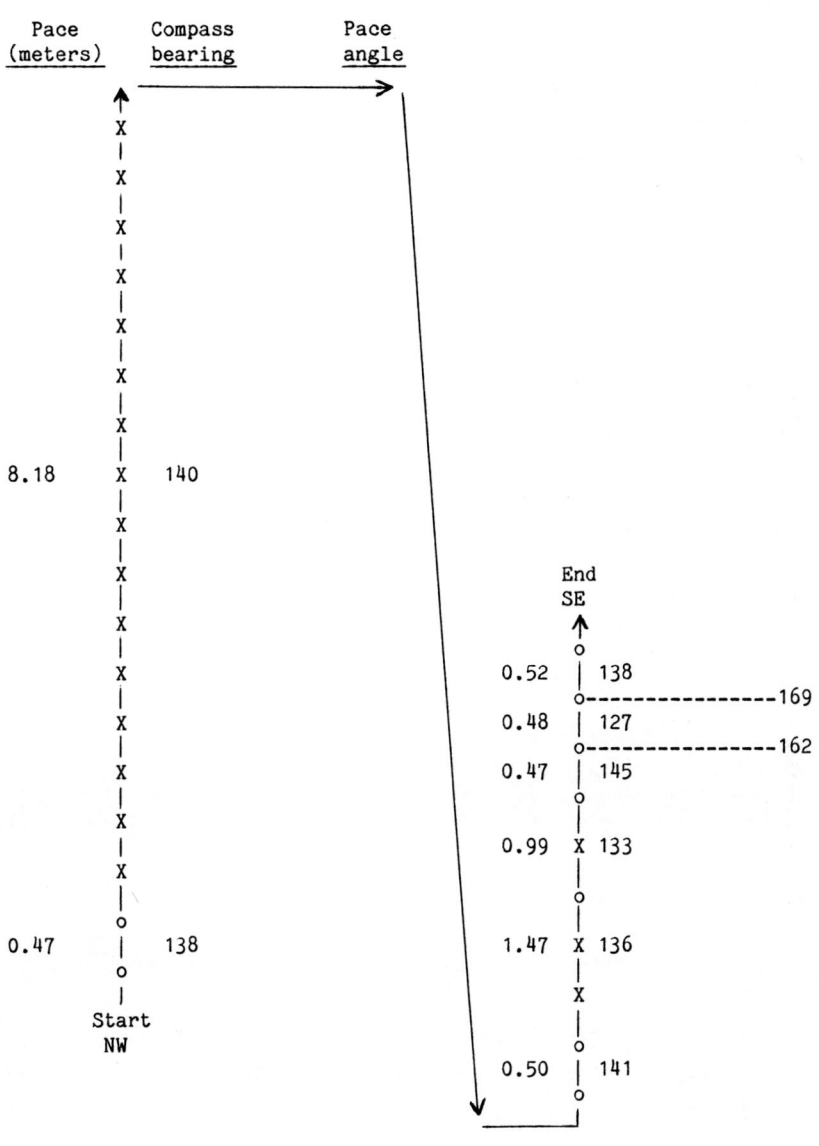

Pace (meters)	Compass bearing	Pace angle

8.18   X   140

0.47   138

Start
NW

End
SE

0.52   138
       169
0.48   127
       162
0.47   145

0.99   X   133

1.47   X   136

0.50   141

9 tracks   X track missing   speed: 2.5 m/s(=9.0 km/hr=5.3 mi/hr)

TRACKWAY NO.: <u>Gregaripus</u> 13

Pace: 0.50   Stride: 0.99   Pace angle: 161

Trackway trend: 113°

Pace (meters)	Compass bearing	Pace angle
	End SE ↑ o	
0.51	| 105	
	o------------------163	
0.48	| 122	
	o------------------161	
0.51	| 103	
	o------------------157	
0.51	| 126	
	o------------------158	
0.48	| 104	
	o------------------164	
0.51	| 120	
	o------------------165	
0.51	| 105	
	o------------------160	
0.51	| 125	
	o------------------163	
0.48	| 108	
	o	
	|	
0.94	X 109	
	|	
	o	
0.46	| 114	
	o	
	|	
	Start NW	

12 tracks   X track missing   speed: 2.6 m/s(=9.4 km/hr=5.6 mi/hr)

TRACKWAY NO.: <u>Gregaripus</u> 14

Pace: 0.51   Stride: 1.02   Pace angle: 169°

Trackway: trend 113°

```
Pace          Compass        Pace
(meters)      bearing        angle

              End
              SE
              ↑
              o
              |
1.02          X    105
              |
              o
0.51          |    101
              o---------------- 165
0.48          |    116
              o
              |
0.99          X    111
              |
              o
0.53          |    101
              o---------------- 163
0.53          |    118
              o---------------- 177
0.53          |    115
              o---------------- 175
0.51          |    120
              o---------------- 165
0.48          |    105
              o---------------- 169
0.53          |    116
              o---------------- 173
0.51          |    109
              o---------------- 165
0.53          |    124
              o---------------- 167
0.48          |    111
              o
              |
0.99          X    123
              |
              o
              |
              Start
              NW
```

15 tracks   X track missing   speed: 2.8 m/s(=9.9 km/hr=5.9 mi/hr)

TRACKWAY NO.: <u>Gregaripus</u> 15

Pace: 0.45   Stride: 0.88   Pace angle: 157°

Trackway trend: 111°


```
    Pace        Compass        Pace
  (meters)      bearing        angle

                 End
                 SE
                  ↑
                  o
  0.43            |     90
                  o
                  |
  0.79            X    101
                  |
                  o
  0.51            |    121
                  o
                  |
                  X
                  |
  1.62            X    117
                  |
                  X
                  |
                  o
                  |
  0.79            X    115
                  |
                  o
  0.41            |    101
                  o---------------- 156
  0.43            |    125
                  o---------------- 158
  0.46            |    103
                  o
                  |
                Start
                 NW
```


9 tracks   X track missing   speed: 2.1 m/s(=7.6 km/hr=4.5 mi/hr)

INDEX